I0866364

DUANE JEFFERY

Introductions to Mormon Thought

Edited by Matthew Bowman and Joseph M. Spencer

For a list of books in the series, please see our website at www.press.uillinois.edu.

DUANE JEFFERY

A Mormon Evolutionist

STEVEN L. PECK

UNIVERSITY OF
ILLINOIS PRESS
Urbana, Chicago, and Springfield

Manufactured in the United States of America
1 2 3 4 5 C P 5 4 3 2 1
♾ This book is printed on acid-free paper.

Cataloging data available from the
Library of Congress

ISBN 978-0-252-04991-0 (hardcover)
ISBN 978-0-252-08957-2 (paperback)
ISBN 978-0-252-4907-1 (ebook)

The manufacturer's authorized representative in
the EU for product safety is Mare Nostrum Group
B.V, Doelen 72, 4831 GR Breda, The Netherlands.
Email: gpsr@mare-nostrum.co.uk

To the triumvirate of BYU friends and colleagues who
played a significant role in my undergraduate education
and in my becoming an evolutionary biologist:
Duane Jeffery, Dennis Shiozawa, and Clayton White,
each of whom contributed in significant and individual ways.

Contents

Foreword to the Introductions to Mormon Thought Series

Our purpose in this series is to provide readers with accessible and short introductions to important figures in the intellectual life of the religious movement that traces its origins to the prophetic career of Joseph Smith Jr. With an eye to the many branches of that movement (rather than solely to its largest branch, the Church of Jesus Christ of Latter-day Saints), the series gathers studies of what scholars have long called *Mormon* thought. We define "thought" and "intellectual life," however, quite as broadly as we define "Mormonism." We understand these terms to be inclusive, not simply of formal theological or scholarly work but also of artistic production, devotional writing, institutional influence, political activism, and other nonscholarly pursuits. In short, volumes in the series assess the contributions of men and women who have shaped how those called Mormons in various traditions think about what "Mormonism" is.

We hope that this series marks something of a coming of age of scholarship on this religious tradition. For many years, Mormon studies has focused primarily on historical questions largely of internal interest to the (specifically) Latter-day Saint community. Such historical work has also mainly addressed the nineteenth century. Scholars have accordingly established the key sources for the study of Mormon history and culture, and they have established a broad consensus on many issues surrounding the origins and character of the religious movement. Recent research, however, has pushed academics into the work of comparison, asking larger questions in two key ways. First, recent scholars have approached these topics from a greater variety of disciplines. There has emerged in Mormon studies, in other words, increasing visibility for the disciplines of philosophy, sociology, literary criticism, and media studies, among others. Second, scholars working this field have also begun to consider new topics of study—in particular, gender and sexuality, the status of international Mormonism, and the experience of minority groups within the tradition. We believe the field has thus reached the point where the sort of syntheses these books offer is both possible and needed.

Duane Jeffery, who earned a PhD in genetics from the University of California, Berkeley, might be less well known than many other figures in twentieth-century Mormonism. But his story touches on many of the main currents of Mormon intellectual life in the mid-twentieth century. Central to that story is the question of modernity: To what extent might a traditionally separatist faith accommodate itself to the trappings of modernity in order to find a home in the United States? Jeffery tested this broad question by becoming perhaps the leading advocate for the theory of organic evolution in the Church of Jesus Christ of Latter-day Saints in a period when most church leaders and a good proportion of its membership were hostile to what some scornfully called "Darwinism."

Jeffery joined the faculty at Brigham Young University (BYU) in 1969. He would spend the next forty years, until his retirement in 2009, carrying the flag for the theory of evolution on campus through a myriad of controversies over science, biblical inerrancy, and religious and science education—controversies he himself stoked by publishing a number of articles asserting the compatibility of evolution and faith. This position drew much negative backlash from some church leaders and professors of religion at the university, and Jeffery nearly lost his job over the fracas. Undaunted, he expanded his efforts beyond BYU's campus, joined the board of directors of the National Center for Science Education in 1994, and engaged in heavily publicized debates over the theory of evolution with prominent evangelical Christian creationists like Duane Gish.

In some ways, Jeffery represents the tensions of the church's attempt to both accommodate to American culture while still maintaining a distinctive identity. He was a devout Latter-day Saint who was trained in and accepted modern biological science and advanced the project of the modern university. To Jeffery, embrace of the broad swath of American science was entirely compatible with his identity as a Latter-day Saint. To others, it was a bridge too far. His career became a battleground over what Mormonism might mean.

Matthew Bowman
Joseph M. Spencer

Acknowledgments

I want to thank those who contributed to writing this book. Jeffery, who has been a hesitant contributor, has allowed many hours of interviews. I also want to thank the Jeffery family, especially Kaye Jeffery and Kodi Jeffery, who have participated in many of the interviews and have been an immense help in recalling many of the details used for this book.

Clayton White, also a friend and colleague, was equally responsible for bringing the first evolution class to BYU along with Jeffery and provided an interview with vital perspectives about how that class came to be.

The most critical person for this work was my research assistant, Beck Seamons, whose support was provided by funds from BYU's Maxwell Institute. His dogged work and research provided critical assistance in making this work possible. He dug in the Jeffery archive at the University of Utah, read through multiple drafts, scoured the archives, and uncovered many hard-to-find resources.

I would also like to thank my research assistant, Ethan Adams, who helped track down important resources and offered perspectives that added to the value of the work. T. Benjamin Spackman gave me early access to his dissertation about the leakage of Protestant Fundamentalism into the church and its influence in shaping the Church Education System's anti-evolution leanings. Michael Austin looked at early drafts and made this a more insightful and refined work. Rebecca Roesler made numerous helpful suggestions but especially saw key parallels I missed between the almost simultaneously published Lester Bush paper on race and the priesthood and Jeffery's work on evolution. My editor, Lori Forsyth, was a relentless, insightful editor who helped me turn this into something not only readable but respectable.

Retired professor Willam Bradshaw's interview was critical in understanding the history of teaching evolution at BYU. William Evenson provided firsthand information and a significant number of papers from his work on the *Encyclopedia of Mormonism* evolution entry and gave insights into his own history with evolution at BYU, both as a physics professor and academic vice president in the mid-1980s (coincidently, I was his undergraduate research assistant). Retired political science and philosophy professor Noel B. Reynolds, who also directed the Foundation for Ancient American Studies and brought it under the auspices of BYU, provided key insights into the creation of the *Encyclopedia of Mormonism*. Gary Bergera provided significant help by allowing me transcripts of his extensive interviews with Jeffery. Jana Riess was instrumental with her Next Mormons Project, which helped me see how belief in evolution among members was waxing and waning within the church. Berkeley Library provided copies of Jeffery's correspondence with his PhD adviser, Curt Stern, housed at the library's Curt Stern Collection.

Dean of the College of Life Sciences Laura Bridgewater and my chair of the biology department, Byron Adams, were generous in their support of this project and added to this work through several important conversations. Mike Whiting, who taught evolution for many years, provided key insights into how the DNA sequencing center was opened at BYU in October 1998, and has been a champion of evolution as director of the Monte L. Bean Museum. Seth Bybee, Steve Leavitt, Riely Nelson, and Clint Whipple were incredibly supportive colleagues, and I bounced many of the ideas in the book off their memories and insights.

Colleagues at the Maxwell Institute, former director Spencer Fluhman, and current director J. B. Haws have supported this work by allowing me to be a fellow of the Maxwell Institute for the six years during which it was embedded.

Series editors Matthew Bowman and Joseph Spencer have been constant supporters of this work. An early copy of Bowman's book, *Joseph Fielding Smith: A Mormon Theologian*, in the same series as this one, was crucial to this work. Alison Syring Bassford, the publication editor at the University of Illinois Press, has been a joy to work with, and her support and suggestions have been amazing and important. The students at BYU are some of the finest I have ever taught and remind me daily why this work matters.

With all the help I was given, I still take full responsibility for the work. I am not sure that the supporting institutions at BYU or its sponsor, the Church of Jesus Christ of Latter-day Saints, the biology department, or the Maxwell Institute, or many of the contributors and helpers would agree with how this work has been framed, executed, arranged, or articulated. I was given much good advice I did not take, and this work likely goes astray in many ways because I am not a historian but rather a writer and biologist. My biases, predilections, and mistakes should not be attributed to those who provided funding or helped me on my way, including Jeffery himself. I retained full editorial control of this work throughout and am responsible for all misdeeds.

CHAPTER ONE

Biological Sketch of a Unique Ape

Delta Boy

Contextualizing any organism begins with understanding its place ecologically. What is its environment? Its habitat? What are the forces that shaped its adaptations to the world? What are the dynamics that made this creature what it is? Relevant questions for any life-form, including the subject of this book.

On September 28, 1937, an ape was born into the world who would influence how science was taught and perceived by members of the Church of Jesus Christ of Latter-day Saints, particularly the science of organic evolution. The primate was a member of the species *Homo sapiens*, a widespread species that evolved on the plains of Africa and spread patchily to every continent, including Antarctica. Of interest here is a small group of this lineage that inhabited a parcel of central Utah on traditional Native American lands occupied by the Goshute, Paiute, and Ute peoples, as well as other speakers of the Western and Central Numic branch of the Uto-Aztecan language family but currently primarily inhabited by Mormon pioneer colonizers who renamed it Delta, Utah. The story of future Brigham Young University professor of evolution Duane Jeffery, and his influence on Latter-day Saint (LDS) thought, begins here.

As is true of many who grow up in small-town Utah, a sense of destiny can be discerned in the Darwinian outlines of what a person survives and how they thrive in the community in which they are embedded. In the northeast corner of Millard County lies the isolated city of Delta, Utah, where Jeffery was raised in the heart of the Sevier Desert along the banks of the Sevier River. Ancient shale formations can be found at nearby Antelope Springs, well known for trilobite fossils—a fitting place for this future LDS evolutionary biologist to find his footing. In 1950, the community had a population of about two thousand souls engaged in agriculture, dairy farming, mineral extraction, and other activities. Delta is also notorious as the site of the Topaz Relocation Center, an internment camp constructed to imprison Japanese Americans during World War II (1942–45), who were falsely believed to be a security risk.

A page from the *Millard County Chronicle*, April 13, 1950, gives a snapshot of the community where Jeffery was raised. A photograph on one of the newspaper's pages shows Jeffery's Sunday school class, newly returned from a trip to Salt Lake City to do baptisms for the dead (a Mormon practice in which living church members are baptized by proxy for their unbaptized ancestors). Jeffery is in the second row, fifth from the left. On the same page, the *Millard County Chronicle*'s ads and articles give a sense of what matters to the community: ads feature suits and ties for men, shoes for babies; buttons that will presumably enhance women's dresses and help them "to catch a beaux [*sic*]!" or perhaps an art-deco kitchen table to add style to a breakfast nook. Articles on the page are various: (1) a list of where Delta citizens have traveled; (2) a close win for the high school baseball team against the Hinkley Mustangs; (3) a navy recruitment notice; and (4) the Fine Arts Club, announcing that it will host an art revue. The paper depicts a city actively engaged in community building, where all the little things that occupy people's lives and through which they find meaning and comfort are on display in the social activities of the citizens.

During his youth, Jeffery's name often emerges in the *Millard County Chronicle*, suggesting a community in which even the activities and achievements of children matter. Jeffery is named as one achieving his Eagle Scout award, then a Silver Palm (fifteen merit badges beyond his Eagle). He is listed as a pallbearer for a young man killed while working with horses. He often sang, both participating in choirs and performing solos. For example, he performed in a vocal recital singing "Friend of Mine" and sang in a thousand-person choir in Salt Lake City. There are several notices of him singing in sacrament meetings and with various choirs. From the paper, it appears he lived an active and full life, participating in Boys' State, wrestling, playing tennis on the high school team, graduating from seminary, and playing an active role in the funerals of friends and family. In short,

Jeffery was meaningfully invested in the rhythms and rituals of this very Mormon community, and this small-town community was invested in him.

In February 1951, the *Millard County Chronicle* records, "Duane Jeffery, son of Mr. and Mrs. Eldro Jeffery, is at the Delta hospital as a medical patient." Although not mentioned in the newspaper, the disease was the mumps, which caused swollen testes and brain encephalitis. At the time, this was a potentially debilitating disease that left many of the afflicted sterile, with some suffering severe mental impairment or even death.

When Jeffery's temperature reached 105° (40.5°), his parents called on Stake Patriarch Charles Woodbury for a blessing. A patriarch in the LDS Church administers vocal blessings of inspired advice and direction to help one get a sense of the blessings the Lord has in store during your life. In this case, it was a blessing for the "sick and afflicted," which differs from the patriarchal blessing. The patriarch told Jeffery that whether the olive oil was used externally or internally, it would have a healing effect. This is unusual. Consecrated oil is almost always used for administrations externally on the body, and later, as the twentieth century progressed, only the head was anointed. After being up with him for several nights, his mother reminded him of what the stake patriarch had said and asked if he would like to take a teaspoon of the holy oil. He agreed, and his mother spooned it into his mouth. The following day, his fever broke, and he began his recovery. Jeffery considered this healing miraculous, especially as it became apparent that his mental functioning was intact, as was his fertility; he later fathered three children.[1]

Another strange, inexplicable event occurred in his teens, something that to Jeffery was indicative of divine protection by powers from above. While driving a tractor, he took a corner too fast, and the machine overturned. He was thrown from the tractor's seat and landed in front of one of the large wheels, which was still spinning wickedly. He hung on to one of the tractor lights as the colossal wheel turned, ready to grind him into oblivion. Suddenly, as his frantic attempt to hang on failed, the tractor engine stopped. He crawled into the seat and found the key had been turned off. No one else was there to do that, and to this day, he has no explanation other than that there was someone or something that wanted him to live.[2]

Jeffery was raised in a caring and active community where he counted as a valued participant, giving him a sense of place and assurance that he belonged in this world. Thus prepared, he followed his older brother Larry north to Utah State University (USU). His brother, a year older, was studying agriculture, and Jeffery joined him in a small basement apartment. He recalls that at that time, they did not understand the workings of academia and found small anomalies puzzling, like calling their teachers "doctors." They had grown up understanding that such honorifics connoted a medical practitioner, not a teacher.

Utah State and Mission to Canada

After two years of studying wildlife management at USU and thinking of becoming a game warden, Jeffery's thoughts turned to missionary service. Like many young men (and a few women) of that era, his first foray into the wider world was his launch as a missionary for the Church of Jesus Christ of Latter-day Saints, in his case destined for the Western Canadian Mission. While serving in Moose Jaw, Saskatchewan, he was made a branch president (leader of a small congregation). As he tells it, "I was brought all the way out from the Vancouver area to be the branch president, and my mission president was an old cattleman from Idaho. . . . He set me apart as branch president, took his hand off [my head], extended his hand for the usual congratulations, and said, 'Congratulations Elder, you are now president of the craziest damn branch in the Church!'"[3]

He was well liked as branch president and got along great with the members. His mission proved to be a practicum for his future in exploring controversial topics and attitudes among the members of the church. Two events became the catalyst for Elder Jeffery being actively engaged in examining complex church issues, a practice destined to influence much of his life, especially in responding to people struggling in the church, as well as the tensions he found working with its leaders and General Authorities.

A Taste for Controversy

In 1958, the LDS French Mission faced a crisis when nine missionaries were excommunicated for apostasy, having become embroiled in promoting polygamy, contradicting the teachings of the church.[4] Many ended up affiliating with Joel LeBaron's break-off polygamist sect, the Church of the Firstborn. One of the participants in the apostasy, David Shore, a missionary who had contacted the leaders of the Church of the Firstborn, was from Alberta, Canada. Because of this, Jeffery was asked by his mission president to watch for apostate activity. He did not find any while he was in Canada, but he was disturbed that such things were infecting the members of the church with what he saw as false doctrine. One night in prayer, he promised God that "if I ever got a chance to kick this stuff in the head—this is my farm boy talk—I would do it."[5]

Jeffery was given the chance to finally confront the LeBaron group when he returned to the States. As he was settling back into college, he was asked by his stake president to investigate the ways the Church of the Firstborn articulated their position on polygamy. Jeffery began reading their literature and visiting members of the group to get a sense of what they were about. Some of those he encountered scared him a bit. After an interview with the notorious polygamist leader Ervil LeBaron, Jeffery recounted, "My wife and I met with him once in Salt Lake, and he regaled me with the idea that if they had had five hundred men of the right quality, they could take over the world. And the implications were very clear: 'You come cast in with us, you'll be one of the big five hundred,' you know? And I said, 'Well, how many men of that quality do you think you've got now?' He said, 'Oh, probably got one hundred twenty.'"[6]

Impressed with his work in refuting the claims of the polygamous groups, Mark E. Petersen (b. 1900), member of the Quorum of the Twelve Apostles from 1944 to his death in 1984, approached Jeffery about writing a response to the LeBaron group who were making inroads in attracting members in the mainstream LDS Church. He was told that the vocal firebrand Bruce R. McConkie, then a member of the First Council of the Seventy, had attempted a response to the group, which had backfired spectacularly. Jeffery writes, "In about 1961, I was corresponding with Elder Petersen and asking for help because I needed help! And he was the Church's man on dissident groups, apostate groups as they were known. . . . He sent me a paper . . . by Elder Bruce R. McConkie with three titles: '*How to Start a Cult*,' or '*Cultism as Practiced by the So-Called Church of the Firstborn of the Fulness of Times, Analyzed, Explained and Interpreted, as also Dissected, Divellicated, Whacked-up, Smithereened, Mangled and Decimated*,' or '*An Essay Showing Where All Good Cultists Go*.' This was somewhere between fifty and sixty pages, and the first eight or ten pages are some of the most brilliant and biting satire I have ever seen in my life." The young Jeffery was not impressed and felt that all the Seventy's attempts to address the

LeBaron faction reflected condescension, instantiated in much of his mockery and dismissal of the group: "He had not bothered to do homework."[7] The Church of the Firstborn had a magazine called the *Ensign* (not to be confused with the LDS magazine of the same name published between January 1971 and December 2020). The magazine spent the next four issues carefully and effectively dismantling McConkie's arguments, or as Jeffery put it, "they gutted him."

Considering the influence McConkie would have against the science of organic evolution during much of Jeffery's career, it is easy to see that some of Jeffery's early interactions with McConkie decreased his deference for the intellectual arguments McConkie later offered. Jeffery took away from these early interactions a sense that leaders like McConkie were very human, with strengths and weaknesses, and were thereby encrusted with the foibles that plague us all. This realization would allow Jeffery throughout his career to grow a commitment to weighing arguments and perspectives based on his assessment of someone's arguments vis-à-vis their comportment to the truths as he saw them and not on their position or authority. In this case, McConkie did not fare well on that score.[8]

Returning to USU after his mission, Jeffery signed up for Reserve Officers' Training Course (ROTC) as was required for university students. This had an interesting effect on Jeffery, who struggled with feelings of academic adequacy, being from a tiny town in rural Utah where a college education was not yet common. For the first time in his life, he felt like he was on equal footing with everyone else and found a place of belonging in the military community: "ROTC was a leveling ground, the wealthy fraternity boys, the athletes, everybody was on the same footing in ROTC, and I quickly became a Platoon Leader and Company Commander, and ended up as the Battle Group Commander, the highest-ranking officer on campus."[9]

During his time at USU, he became interested in evolution through a genetics class with USU scientist Eldon Gardner (1909–89), one of the country's leading human geneticists and an LDS bishop who became a role model for Jeffery. Gardner also encouraged Jeffery's interest in evolution through a small pamphlet his professor had written on evolution and the Bible. In the next chapter, we will look at the relationship between the science of organic evolution and certain leaders in the church in more detail, but in brief, there were genuine tensions among leaders in the church about the science. Because Jeffery's work on the Church of the Firstborn was so effective, he soon became the go-to person for answering controversial and complex questions, including those about organic evolution. He said in a 2009 interview,

> I was also in a stake mission presidency at Utah State at that point, and for reasons I don't understand, it seemed like all the tough problems landed on my desk. Students who were concerned about science issues were instantly referred to me. I had picked up a little history and a little interest in Mormon history, and so anybody with those problems got handed over to me.[10]

Jeffery was drawn into the controversy surrounding Darwin's theory through his interest in evolution, sparked by his mentor Gardner and by the writings of one of evolution's most prominent critics in the church, Apostle Joseph Fielding Smith, as a result of his engagement with the apostle's thinking as a university student. During his undergraduate years, when asked about the influence of Fielding Smith's controversial 1954 book, *Man, His Origin and Destiny*,[11] a targeted attack on evolutionary science suggesting the Earth is only six thousand years old, Jeffery commented that among the LDS faculty, it had little influence, but among ordinary members, "it was Joseph Fielding Smith all the way. Evolution was just not to be countenanced at all."[12]

The controversies that would later define much of his work at BYU on evolution appeared as Jeffery confronted questions about the relationship between science and religion. He gives a sense that Fielding Smith's book was creating divisions among the science students on USU's campus. One of his fellow students threw the book across the room and exclaimed, "That's it. If I have to believe this stuff, I'm through with the Church!"[13]

Jeffery felt that Fielding Smith's work was making his work as a stake missionary very difficult. While among engineers and students from the humanities they were making converts, the book had poisoned his efforts among biologists, and "none were interested very long."[14]

We will examine the cost that Fielding Smith did to the intellectual climate of the church and its reputation among scientists in the next chapter, but note here that the damage started almost immediately after the publication of Fielding Smith's book, which was perceived among scientists as vehemently both anti-evolution and anti-science.

Jeffery found support for a view of organic evolution that was compatible with Mormonism through LDS writer Nephi Anderson's novel *Dorian.*[15] Anderson quotes extensively from Henry Drummond, a religious thinker whose book *Natural Law in the Spiritual World* was supportive of more scientific approaches and was widely read by the Saints, especially because it was a recommended text for church instruction at the turn of the twentieth century. This work would play a crucial part in Jeffery's integration of science and religion.[16] In an interview with Keith E. Norman published in 2002, Jeffery says this about Drummond's work:

> It was really a spiritual experience for me because Drummond's argument was that the natural laws we see here in the physical world are just extensions of the laws recognized in the religious universe. That resonated greatly with the Mormonism I had been taught.[17]

In these sources, Jeffery found the possibility of a reconciliation between his faith and his growing interest in Darwin's theory of evolution.

Marriage and Starting a Family

In 1959, shortly after he came home from his mission in Canada, Jeffery's old missionary companion Elder James Howard Westover came to see him in Logan and mentioned he had a sister named Kaye at BYU majoring in English.

Shortly after, the Westover family came to Logan to visit, and Jeffery accompanied them on a day trip to Dinosaur National Monument and then drove with them back to Provo. He spent the evening at the Westovers' home and was invited to stay the night. Kaye and Jeffery were sent to tell the kids downstairs to go to bed. They went down to convey the message but failed to return. Wondering what the young couple was up to, some of the adults checked on them but found them only in animated conversation.

This flowered into a romance, and a few months later, the lovebirds were engaged. For LDS, they had a more prolonged engagement than was typical at the time, as Kaye's father was in part concerned that his daughter was only eighteen. Therefore, Jeffery and Kaye made a deal with her parents that they would wait for a year, and Kaye's father would assist Jeffery financially in his USU studies. Jeffery suspected that her dad hoped the long engagement would give the young couple extra time to think through things more carefully.

The extra time did not lead to second thoughts about each other. In 1961, Kaye graduated from BYU in English, and she and Jeffery married that fall.

Jeffery's Formal Scientific Research Begins

After finishing his bachelor's degree in wildlife biology the same year he married, he decided to go on for his master's degree in wildlife ecology. Funded by the U.S. Forest Service, Jeffery's first scientific endeavor was studying the extensive elk range habitat that comprises approximately eight thousand acres in Willow Creek Canyon in the Fish Lake National Forest near Salina in south-central Utah, under the direction of Dr. Fred H. Wagner in the College of Natural Resources. To better understand how to manage elk populations, he created a statistical model looking at vegetation, elk droppings, and other variables to characterize the resources of interest to elk in their struggle for survival.[18]

By the end of his field research on elk, Jeffery's interest in wildlife science was waning as his fascination with the science of genetics was increasing. One day as he was finishing his master's degree, he walked into the office of his genetics professor, Eldon Gardner, and asked, "Where's the best genetics PhD program?" Gardner, whose work eventually comprised over three hundred papers on the subject, told him it was Berkeley, where Curt Stern was doing some of the finest work in genetics in the world. Jeffery knew little about the place, but after some investigation he sent in an application to study genetics at Berkeley.

Following the ROTC support during his undergraduate education, there was an obligation to serve in the army after he finished his schooling. Although he could have gotten an educational deferment, he decided to join up. He had done well in ROTC, gaining confidence, and discovering he had a knack for leadership. And the life of an officer in the army had some appeal. It was a life of order, and he sensed that it would provide stability and let him pay off his debts—like many young couples, he and Kaye had financial challenges. He decided to tell James Bell, the president of the stake mission, about his decision to serve in the military rather than go to

Berkeley for graduate school in genetics. Bell's sardonic response changed everything:

> Instantly he gave me the worst dressing down I have ever had in my life, just ridiculed me. "You know, yeah, go into the army and holler '1-2-3-4,' any old lady can do that, you big boy!" By the time he got through with that, he said, "Sure, Berkeley will give you an assistantship any old time, you can just turn them down, they'll have it waiting for you when you get back, no problem." When he got through, I was not going to go to the Army if I could help it. So, I applied for a further extension in call to duty, it was granted, and I went to Berkeley.[19]

In 1963, Jeffery finished his master's in wildlife biology and, with his deferral for military service in place, prepared to go to Berkeley to accept a teaching assistantship. Kaye was pregnant with their first child and did not immediately go with Jeffery to California but remained with her sister in Malad, Idaho, where she gave birth to their first daughter, Tammi, in September of that year. Jeffery would not see his new daughter for three weeks after her birth. Their second daughter, Kodi, followed in 1966, and their youngest daughter, Doni, was born in 1968. As well as laying the academic foundation for his future career, Jeffery's graduate studies also formed the backdrop for the births of each of Jeffery and Kaye's children.

Jeffery's Experiences at the University of California, Berkeley

Jeffery arrived at his new adventure at the Department of Zoology and Genetics, University of California, Berkeley without communicating with his PhD adviser, Dr. Curt Stern (1902–81). Jeffery attributes this lack of communication to his naivety about how graduate school works. To his surprise, he learned that his would-be adviser had recently left for a sabbatical. At the time, Berkeley was rated by the American Council of Education as the number one graduate program in the country.[20] Stern was a world leader in genetic research. He had done foundational work in "crossing over," or how genes recombine into new configurations to maintain genetic variation in fruit flies (*Drosophila melanogaster*). When Jeffery joined his lab, he had turned to gene regulation in humans, while still using Drosophila as a *model organism. Curt Stern, a member of the National Academy of

Sciences and the author of a popular textbook on genetics, was one of the most well-respected geneticists in the world. Jeffery arrived intimidated and overwhelmed. In his 1981 interview with Gary Bergera, Jeffery said, "Berkeley <was> both an extremely heady and an extremely frustrating experience."[21] Jeffery, however, would go on to do meaningful work under Stern, completing his dissertation on radiation-induced chromosomal rearrangements that produced several Drosophila fruit fly colonies that exhibited flies carrying rearranged chromosomes with varying degrees of "hairiness." This work was subsequently published in the highly respected professional journal *Genetics*.[22]

His first teaching assistant job on campus was teaching freshman biology, which he enjoyed and which allowed him to patch some holes in his education in wildlife biology. However, during his first semester he signed up for too many classes and did not do well. He did so poorly he feared he might lose his teaching assistantship.

Happily, he not to have worried. When his adviser returned, things settled into a more typical graduate student experience teaching and researching the discipline.[23] He was asked to teach an embryology class and thereby learned the fundamentals of current developmental biology more thoroughly.

In addition to the usual problems of navigating graduate school, church was demanding. While at USU, he had been very busy in callings and working with General Authorities responding to the literature of the Church of the Firstborn, but he began to be concerned about how he might handle a similar workload during graduate school. The first night he arrived in Berkeley, he told Kaye during a call, "You know, we've got to leave this stuff alone, I'm in the big time now, I've left the cow college."[24]

Fate and circumstance had other ideas. Shortly after he arrived in Berkeley, the Church of the Firstborn arrived in the Bay Area and began to distribute handbills on cars parked at the stake conference that Jeffery and his family attended. The local church leaders were concerned. Jeffery relates, "So, when the university ward got it, it created a stir. Who knows anything about this? And I was dumb enough to say, 'Well, I do.' And so, I was asked to give them a little, so I taught priesthood meetings for six weeks on the Church of the Firstborn." Word spread of his 'expertise' on this matter and the Northern California Mission president called him to a special mission to work with members struggling with the Church of the

Firstborn and polygamy. Jeffery told Kaye, "So, we were right back in it again!"[25]

In addition, he was involved on a committee to rewrite the Aaronic Priesthood manuals, which were thought to reflect too much of Wasatch Front culture. LDS pioneers, who established much of the Mountain West culture, often used the stories that emerged from agriculture and small-town communities in their discourse. This colored the ways they spoke about their lives in ways that might not have been shared with those from other backgrounds, like those Saints living in heterogeneous California. So, despite the demands of graduate school, he was again swamped with the demands of church callings. These became even more challenging as he accepted a call as second counselor in the bishopric despite expressing deep concerns about how he would manage the time demands of the calling on top of his graduate work at Berkeley. He was especially concerned considering the number of LDS graduate students at Berkeley who were flunking their prelims and dissertation defenses due to paying more attention to the demands of the church's service requirements than their schooling and thus being forced to abandon their programs—a fate he resisted. Even so, it was part of his Delta, Utah–based faith that one did not turn down callings, so he accepted the call to serve.

During this time, the Berkeley campus was going through the throes of the cultural shifts of the 1960s, which included Vietnam War protests, challenges to cultural mores, and the civil rights movement. Jeffery was largely untouched by most of the foment but struggled with the changes in his personal life. Jeffery's commitment to the church at this time was unimpeachable. The lifelong storm that would descend on Jeffery when he began teaching and writing about the history of evolution in LDS thought after he arrived at BYU, and for which he was seen as a heretic, belies the actuality of his religious praxis.

Jeffery's graduate work included more coursework on organic evolution and its Darwinian roots, necessitating confronting more directly the church's relationship with organic evolution. Evolution lies at the center of all of biology's theoretical underpinning. It is the most vital scientific concept that undergirds the entire discipline. He realized that the historical background of the church and evolution was more complicated than he had been led to believe. Fielding Smith and Bruce R. McConkie's narrative about the diabolical role evolution played in science was incomplete and dangerously distorted from the actual science. Their view represented a biased perspective of both evolution and the discipline's relationship to the church. He began to feel that assisting in creating a more accurate and nuanced view was warranted. He sought help from Ken Godfrey, the director of LDS religious instruction at Stanford.

One night, they stayed up late talking, and Godfrey recognized how deeply Jeffery was fascinated with the science of evolution. A new chapter was about to open in his life, contextualizing much of his engagement with the church, his future employer, BYU, and his place in the history of evolution and the church. During his visit with Godfrey, the director asked Jeffery if he was interested in evolution, saying he had been handed an unknown detailed historical description of a 1931 attempt to settle the church's relationship to evolution among the leading general authorities, like B. H. Roberts and Joseph Fielding Smith. Jeffery was shocked at the 1931 letter and after reading its contents revisited the respected institute director, asking about the letter's authenticity, which Godfrey confirmed. Jeffery began talking to other scientists and found that none of them had heard of the letter either.[26]

By this time, Jeffery had worked with several General Authorities. His interactions with Mark E. Petersen and his engagement with the ill-conceived material that Bruce R. McConkie had written to address the Church of the Firstborn described above gave him different perspective on these leaders than the average member of the church might have a chance to develop. A small event illustrates the complexity of Jeffery's relationship with General Authorities:

> While I was on my mission, we had a visit from Elder McConkie, and then I started to wonder a bit about it, because we had fourteen investigators out to a conference that he addressed, and not one of them would let us back in the door. And so, I started to wonder about the universality of his ability to relate to people. . . . That didn't have heavy effect on me, but it certainly was always in the back of my mind on a number of issues.[27]

Here he once again suggests that Elder McConkie was so brusque and off-putting that once people were exposed to him, investigators to the church refused to come back. As a result of such experiences, publications that attacked evolution like *Mormon Doctrine* held less sway in his mind.

His work with Petersen and other leaders of the church, together with his experience growing up in a time when relationships with central leaders were more accessible, led him to see them as respected but flawed individuals and made him feel free to communicate, write, and speak to them as fellow citizens of the kingdom. He saw them as more approachable, human, inclined to biases and mistakes, and deserving of respect for their positions of authority. The hedges that separate church leaders from the hoi polloi of church members were much thinner then than they are now.

Much of his later work in promoting and illuminating the history of the church and evolution required recognizing that several leaders of the church had made mistakes in promoting anti-evolutionary perspectives to the Saints. Challenging the leaders of the church on these issues was going to become one of the main barriers in bringing to light the ways evolution and his faith tradition were compatible. A brief look at some of the disappointments in his experiences with certain General Authorities offers a sense of why he had few compunctions contradicting fully human men with influential positions in the church. It must be remembered that, like his friend and contemporary Eugene England, Jeffery grew up in a very different world than subsequent generations will experience. Kristine Haglund captures this world nicely in her biography of Eugene England where she describes the familiarity of members of the church with General Authorities who would come and stay overnight with local families and even bless their crops gave them a more intimate view of who they were as ordinary people.[28]

Like England, Jeffery carried on extensive correspondence with General Authorities (currently forbidden at BYU because of the demands on the time of General Authorities who run the global church). He felt comfortable challenging them and asking their opinion on matters, but he always focused on seeking the truth as he saw it—including on those questions informed by science. Although he respected their position, the deference that England exhibited in his well-documented confrontation with certain General Authorities[29] was not a part of Jeffery's nature. His science training that inclined him toward recognizing empirical evidence from nature acted as a check on how much weight to apportion to others' opinions, no matter their church position. He could respect General Authorities but did not suffer what he saw as recklessness in embracing ideas pulled from biblical readings that contradicted the received view of evolutionary science.

It is important not to read too much into what might appear to contemporary church members as his disrespect of some leaders of the church. In her book on Eugene England, Haglund describes the shift that happened during the period when both Eugene England and Duane Jeffery grew up, moving away from the possibility of intimate relationships between members and leaders as the church moved from a more regionally circumscribed to a worldwide organization. She notes that in the decades that both Jeffery and England grew up in rural areas of Utah, there was much more intimacy between ordinary members in these communities than today with Apostles and Prophets visiting to speak in Stake Conferences and to council local members. Almost always they stayed and took meals with local families. This gave members greater access and familiarity than contemporary members have giving them a more human touch. This explanation aligns with Jeffery's descriptions of the human aspects of General Authorities as reflected in his interactions and relationships with them over the years.[30]

An early example of Jeffery confronting General Authorities as human actors suggests how these interactions contributed to the dismantling of his sense of their infallibility. He had been working with and became friends with a repentant husband and wife who had been excommunicated from the Church of Jesus Christ of Latter-day Saints for joining the Church of the Firstborn. Having a change of heart, the couple said they wished to join the church again. Because it was a high-profile case of former members returning from an "apostate" group, they had to be interviewed by a General Authority. In this case, it was Mark E. Petersen, who had previously directed the efforts to limit the influence of the Church of the Firstborn and who had overseen Jeffery's pamphleteering against the polygamist organization. Petersen seemed to be the point man for addressing other Mormon groups that had dissented from the larger Brighamite branch of the church. Jeffery waited for word of the interview and about his friends' return to the faith.

Finally, Jeffery wrote his friends, who reported, "Brother Petersen came down, and we had a very congenial chat, but everything went dead." It turned out that after talking with them, Petersen claimed that the couple had failed the interview.

Jeffery continues,

> I wrote to LeGrand Richards [an Apostle of the church]. I was brassy enough to do this kind of stuff in those years. You know, there was not quite the heavy counsel that you just don't write to General Authorities

> that has been for the last twenty-five or so years. And probably because of guys like me, I guess. But I wrote to LeGrand Richards and he wrote me back about a two-page letter, and he quoted about half a page from Elder Petersen's report of the meeting dealing with my friend, so let's call them "Jack" and "Jill."
>
> And man, that description did not describe anybody I ever knew. . . . I couldn't relate at all. And Brother Petersen [reportedly] said, "You can see, we clearly cannot have unsavory people like this in the Church." I mean it was sexual profligacy and all sorts of stuff. The usual thing we used to explain anybody that leaves the Church, you know, they're always sexual reprobates, whether they are or not, but that's our excuse.
>
> At any rate, I called up "Jack," . . . he was just stunned. He says, "We didn't talk about any of that stuff. That's a total fabrication." From Elder Petersen reporting on Jack and Jill! And I said, "Oh, this is great. You and me against a member of the Quorum of the Twelve," and he said, "No, our stake president was there, he'll back us up." . . . So, I called the stake president and I read the letter to him over the phone, and he said, "I don't believe you." He said, "That has nothing to do with what we talked about. I'm coming up for General Conference. Could I stop into your office?"[31]

The matter was cleared up, and the couple was baptized. Jeffery had no qualms about using his connections to one Apostle to challenge the actions of another whom he felt had made errors or even strayed from the facts of the matter.

Petersen and Benson would become two of the most vocal antagonists against Jeffery's evolutionary efforts, including trying to get him fired from BYU. This example illustrates the complexity of his relationship with General Authorities, who both disappointed and supported him at various times and helped him see them as flawed individuals who, despite their other admirable qualities, were capable of mistakes and errors.

CHAPTER TWO

Origin Stories

Evolution and Religion in Conflict

In 1968, Jeffery received a phone call from Lester Allen, the chair of the zoology department at BYU. At the time, Jeffery had a meager opinion of BYU and had no intention of going there, but Lester said, "We'd like to offer you a job." Surprised and somewhat shaken, Jeffery replied, "You don't know anything about me." Lester replied, "Well, aren't you just finishing up a degree with Curt Stern?"

"Well, yeah," Jeffery answered.

Allen asked, "Aren't you in the bishopric of the university ward down there?"

"Well, yeah," Jeffery said.

"Well, it sounds to me like we've done our homework. We'd like to offer you a job. What do you say?"[1]

Jeffery checked all the boxes; he served as a church leader and had indisputable scientific clout through his Berkeley education.

Jeffery had not resolved the question of whether he had a military obligation after his PhD and expressed those concerns, but Allen suggested he come out for a visit, and Jeffery took him up on the invitation. During the visit, he learned that although the groundbreaking had taken place for the new Widtsoe Building, which would house the life science labs, Jeffery would have to set up a temporary lab in less-than-ideal conditions. Not an optimal situation.

Nevertheless, Jeffery was worried about his family and their growing financial needs. He now had three children and had begun to question the wisdom of his plan to join the Army Infantry Medical Corps in the shadow of the war in Southeast Asia. Vietnam added to his misgivings about serving in the military. He had a friend who had engaged in combat there and suffered what we now call severe post-traumatic stress disorder. The man had changed so much that Jeffery was concerned about what serving in Vietnam would do to him. Once again, fate intervened. At a military exam to assess his fitness to serve, it turned out that stomach ulcers he had developed on his mission were still unresolved, and he was ineligible for a commission as an officer. The problem was solved, and he decided to come to BYU.

In February 1969, Jeffery arrived at BYU, about half a year before he planned to finish his dissertation. For promising graduate students, it was not uncommon to start a job before finishing to lock in the student to a position that might be hard to fill. He would later call the decision to arrive early the stupidest thing he had ever done.[2] He came partly because BYU desperately wanted a geneticist working there for its zoology program. Jeffery still had some data to gather and analyze to finish his PhD, but the temporary lab space he was allotted, in some old houses where the Marriott Center now stands, was not conducive to his scientific work in genetics, especially when compared with the facilities he enjoyed at Berkeley. To make matters worse, after a few months, the houses were torn down, and he was moved to the old BYU high school, where he had to set up his lab all over again.[3]

These were difficult times, weighted with many pressures, including the needs of a young family and an enormous amount of work setting up his lab (twice), which included rearing untold numbers of fruit flies (*Drosophila melanogaster*), examining their genetics, and analyzing the resulting data. In all this, he was "getting three, four hours of sleep a night" because he had to redo some of his experiments up to seven times to get to his final datasets.[4] Finishing his dissertation at BYU took him two and a half years, whereas had he stayed in California with the more well-established and better equipped genetics labs at Berkeley, he would likely have been able to finish much earlier.

Conditions at BYU When Jeffery Joined the Faculty

Jeffery arrived at BYU near the end of controversial Ernest L. Wilkinson's term as president. Flawed but brilliant and driven, Wilkinson had two goals for BYU. First, he hoped to increase the size of the student body. When he took over the school's presidency in 1951, only two thousand students were attending BYU, but by the end of his tenure, he had increased that number to around twenty thousand students. Wilkinson's second goal was to increase the academic reputation of the university. A desire to add heft to the university's prestige also may have been one of the reasons Jeffery was hired.[5]

While Wilkinson successfully increased the school's enrollment, he also created "a darker legacy of coercion and distrust"[6] in response to his perception and fear that many students were being caught up in the countercultural response to and critique of "establishment" values. To combat what he saw as student embrace of these liberal, and therefore dangerous, politics, he began to use the honor code as a standard of righteousness both as a means of setting expectations for student behavior and as a proxy measure to identify students who were not keeping the commandments. In his last six years as president, he became obsessed with identifying students and faculty who were not living up to his perceptions of church standards. He conflated his own strong bias against political liberalism with his take on gospel requirements. As such, he worked tirelessly to ferret out those who sympathized with the civil rights movement, interracial marriage, communism, homosexuality, and other political hot-button topics.

Wilkinson was certainly not alone in his political views among the church leaders of the time. Gregory Prince, over the course of his book on the life of David O. McKay, shows that many of Wilkinson's views were shared by David O. McKay, Ezra Taft Benson, Mark E. Petersen, and other leaders of the church who have figured in the Jeffery story. (A prominent counterexample with more liberal political leanings was Apostle Hugh B. Brown, a counselor in McKay's First Presidency.) These leaders had significant differences, and their views were complex and nuanced.[7]

Wilkinson's obsession with identifying errant students also extended to figuring out faculty not faithful to his narrow version of the church, which conflated the gospel with a political conservatism that eschewed communism and socialism and embraced right-leaning causes (including resistance to civil rights legislation for descendants of the nation's enslaved Black population).[8] In his zeal to enforce purity, he used students as spies and modeled his attempts to identify subversion on national espionage work. This included trying to root out undesirables by requesting permission from President McKay to hire a former Federal Bureau of Investigation agent to head up his surveillance of students and faculty, a request that McKay approved.[9] Wilkinson's efforts began the abatement of freedom in ways that still reverberate through BYU's faculty.[10]

While Jeffery was being wooed to come to BYU, LDS ornithologist Clayton White was also being recruited to join the life science faculty. Clayton White and Duane Jeffery began a collaboration to bring the teaching of evolutionary science back to BYU where it had not been systematically taught for nearly sixty years.

Clayton White

Clayton White, a PhD raptor specialist out of the University of Utah's biology program, was doing landmark, cutting-edge research at Cornell University's Ornithology Lab at the forefront of one of the most pressing ecological problems of the latter half of the twentieth century. Environmental researcher Rachel Carson's seminal book *Silent Spring* was published in 1962 to worldwide acclaim (and notoriety). In her data-rich tome, she carefully and convincingly demonstrated that pesticides used to control insects were implicated in the disappearance of birds, beneficial insects, and other earthlings vulnerable to these toxins being sprayed to protect crops and livestock. One of the pesticides was an organic phosphate called DDT (dichlorodiphenyltrichloroethane), a broad-spectrum pesticide used for controlling insect and mite pests on agricultural crops, ticks on livestock, and mosquitoes in urban areas.

As Carson's research suggested, Northern and European populations of the peregrine falcon, until that time quite common, began to crash. Researchers discovered that their eggshells were thinning to the point that even the weight of the nesting parents would break the eggs. In addition, hormonal changes in the birds were creating conditions that curtailed natural maternal instincts such that mother falcons would abandon their chicks, even if the eggs happened to hatch.

Clayton White was asked to look at and document the viability of Alaskan peregrine falcon populations. This work attracted the attention of BYU administrators who were actively trying to bring the school's science departments up to national educational standards. To ascertain White's fitness to join the faculty at BYU, they invited him to come to campus to present his research.

White reported that he blew his first interview because after discussing what he could do for BYU as a faculty member, he dared to ask what BYU could do for him. Apparently, the administrators felt the privilege of teaching at the Lord's university should be enough for anyone to take the job, so they rescinded their offer. A year later, having not found anyone suitable for the needs of BYU's life science programs, they were again knocking on the door of one of the world's premier peregrine falcon researchers and offering him a job in BYU's zoology department. Despite having two offers elsewhere, including at Cornell University, White accepted.[11]

BYU's First Undergraduate Evolution Class in Sixty Years

In exploring the science of evolution, Jeffery found a kindred spirit in White, and they determined to bring an evolution class to BYU's zoology department. Evolution had not been taught formally on the BYU campus since the 1911 firing of Peterson brothers, Joseph and Henry, and the brothers Ralph and William Chamberlin. While this story has been told in many places, a brief overview is warranted and will be described briefly below.

When Jeffery and White arrived on the scene in 1969, evolution had been discussed superficially and without fanfare by individual professors, but no formal undergraduate class had been offered in modern times. The one graduate course that had been held was taught by botanist Howard Stutz to four or five students.[12] During Jeffery's interview for the BYU position, the upper-level administrator admitted, "Well, I think you know that BYU's biology department is not highly respected." Jeffery answered, "Yeah, yeah, I know," then asked, "Do you have any idea why?" The interviewer acknowledged that it was because evolution was not taught. "Well, if I were to come here, I would propose a course in evolution. How would it be received?" The interviewer replied, "Like any other course, it would be evaluated on its merits. And if it's got good merits, it would be approved. It has to go to the board of trustees, but it'll get a fair hearing."[13]

This event was pivotal in the history of evolution being taught at BYU, and it is worth looking at how Jeffery remembers the event in detail:

> But in the Fall of '69 or so, Clayton White came, and he and I put together a proposal for a class in evolution and pitched it at the senior level, which was where it would seem to be best received, and we . . . could rely on their background. So, we pitched it there. I was surprised at the

curriculum committee here in our college—the head of that committee just laid it out to us and said, "Well, we talked about this class, but do other universities teach anything like this?" It just blew me away. And I said, "Well of course they do! We're asking one dinky class! They teach them in the handful!" And he says, "Well, would you go to catalogs in the other universities and kind of bring to us what they offer in the way of evolution?"

So, we came back with something like 34 courses from Harvard and 37 or 38 from Kansas, and you know, just laid it out. I couldn't believe it, that the head of a biology curriculum committee was asking that question. But that was the end of that. We made our point. And so, we put it in, and we were called one day by [Academic Vice President and Honors Director] Robert K. Thomas, who said, "I need to let you know that your course has been approved and you can teach it this coming Fall," Fall of '71. And we said, "Okay, we will do that." And he said, "But I need to tell you a little more about it here. I took it up with [future president of the church, then Apostle] Harold B. Lee," who was the power behind the throne in those days, "and he looked at it and said, 'Oh, I know they need this, but I don't want to wrestle with Brother Benson in a trustees meeting on it. Why don't I just approve it, and you tell those brethren to teach the most rigorous and best quality course of which they're capable, just don't get on any bandwagon and beat the Church with it.'"[14]

In addition to the school not matching the standard requirements of most life science programs, there was direct evidence that BYU students were harmed by not having a course in evolution for their biology degrees. Soon after he arrived at BYU, Clayton White was put on the Dental Committee and asked to help shepherd the students through the processes of getting into dental school. In particular, he discovered that one of the best schools had not even acknowledged several BYU applications. Clayton contacted some of the schools and after several failed attempts, an officer from Northwestern University finally admitted to him, "Well, you know, Dr. White, we look at these applications and we look at the background, and we don't see any course in evolution at Brigham Young University. So, we even wonder how well-trained your students are in the biological sciences, and that's why we don't accept them."[15]

So, with Lee's approval, Duane Jeffery and Clayton White taught the first undergraduate course in evolution in the zoology department at BYU in Fall Semester 1971, after nearly sixty years of absence. Listed in the BYU catalog in 1972 as Zoology 404, Comparative Evolutionary Theory, the course was described as "a critical study of principles of organic evolution and its impact on modern thought." In 1981 it would become a required capstone class for zoology majors.[16]

This was a busy time for Jeffery because he was putting in long hours at the complicated and physically taxing work of setting up his genetics laboratory. Even during this busy time, his thoughts were captivated with the history of evolution and its place in the theology and doctrine of the church. Robert A. Rees, editor of *Dialogue: A Journal of Mormon Thought*, had conceived of an issue devoted to the topic of science and religion and tasked BYU zoology assistant professor James L. Farmer with putting it together. Because of his known interest in the subject, Jeffery was asked to contribute an article on evolution and the church.[17] He had been framing such an article and exploring its required research since his time at Berkeley, so he jumped at the chance to contribute to this issue and articulate his research on the subject. Titled "Seers, Savants, and Evolution: The Uncomfortable Interface," this landmark paper was published in the autumn/winter issue of 1973.

Immediately after its publication, he left on sabbatical to the University of Hawai'i, which precipitated rumors that he had been fired for his pro-evolution stances. Nothing could have been further from the truth. In chapter 3, we explore the reaction to his paper in some detail. Next, however, to put Jeffery's work in context, we lay some groundwork to explore what evolution is (especially as it is often misunderstood). A sketch of the relationship between organic evolution and several influential leaders of the Church of Jesus Christ of Latter-day Saints follows.

Problems in How Evolution Had Been Taught to Members of the Church

Many of the especially vocal leaders, such as Joseph Fielding Smith, Bruce R. McConkie, Ezra Taft Benson, Boyd K. Packer, and others, embraced Protestant Fundamentalist creationist ideas that established an atmosphere of anti-science, one that would pervade Church Education for much of the late twentieth century and beyond. This anti-science stance appeared in conference talks, manuals, and other educational resources for instructing the youth of the church. This difference in opinion between certain strains of LDS religious thought and science created animosity between the life sciences and religious education programs. This tension became so acrimonious that shortly after Jeffery's arrival, BYU president Dallin H. Oaks attempted to mitigate the fracas by setting up a series of seminars to start a conversation between scientists and religious educators about evolution and science that included Jeffery as a participant. Unfortunately, in the end, these efforts would fail. Thereby, many of Jeffery's most vehement detractors in the early years of his tenure at BYU came not only from BYU's College of Religious Education but also from some of the leaders found among the church's General Authorities. Understanding these tensions is important for contextualizing how the teaching of evolution emerged at BYU.

Armand Mauss's conception of cycles of "assimilation" and "retrenchment" within the church is a valuable way to contextualize the story of evolution in the church.[18] In his book *The Angel and the Beehive*, Mauss identifies a pattern of oscillation that characterizes organizations that attempt to be recognized as a part of mainstream society but that are at odds with the beliefs and behaviors of that society. This framework becomes a good lens through which to view the church's stance toward evolution vis-à-vis its religious commitments and priorities.

As Catholic theologian John Haught argues, "any thoughts we may have about God after the life and work of Charles Darwin can hardly remain the same as before. Evolutionary science has changed our understanding of the world dramatically, and so any sense we may have of a God who creates and cares for this world must take into account what Darwin and his followers have told us about it."[19] Darwinism necessitated a new image of the cosmos and created a division between two dominant poles—those willing to accommodate what science was discovering about life on Earth and those who turned to Protestant Fundamentalism to stave off the fear of the troubling theological implications of evolution as a mechanism deeply embedded in the Creation.[20] The range of thought between these two poles is too nuanced and complex to deal with in a short introduction, but this work will at least establish the boundaries of this dichotomy while acknowledging there were and still are a range of responses to Darwinian evolution. As will be demonstrated in the book, the power of Jeffery's influence and the impact and flow of his work within this debate was profound.

While at Berkeley, Jeffery began to accumulate evidence that much of what was assumed and taught by and to members of the church about evolution paid little attention to modern science, especially biology. In much of the discussion among religious leaders, there was little truck with the actual science of evolution, and it relied on outdated information circulated in the emerging creationist literature. Jeffery was especially concerned that material used in the church's educational system was disastrously misinformed and flawed about science and that many of the perspectives and conclusions of those working in religious education were unwarranted. Much of the culture of church education borrowed from Protestant fundamentalist teaching that God created the Earth on the order of six thousand years ago and ideas imbued with fanciful theological speculation about race, gender, and sexuality. It also ranged over many of the scientific ethical discussions going on in the life sciences about everything from the implications of certain medical practices, including abortion, to caring for the environment. To explore this, a short detour to examine the science of organic evolution is useful.

What Is Evolution?

The idea that species could change and even go extinct and that life was a dynamic process in which variations became the grist for substantial changes in form and function developed slowly. Evolution's roots in classical philosophical thought began with thinkers from antiquity like Plato, Aristotle, Xenophon, and Lucretius. However, several early Christian fathers problematized this view by incorporating Neoplatonist ideas, including creation ex nihilo, enforcing the idea of a "great chain of being," which embraces notions of species fixity and an ordered hierarchy among created things that stretched from God at the top to the earthy mineral world on the bottom. These early Christian thinkers turned the Creation into a doctrine drawing heavily on Greek philosophy while ignoring its Levantine roots, thereby creating theological conditions steeped in tensions that would eventually emerge between Enlightenment biology and religion.[21]

Biology in the sixteenth and seventeenth centuries became a dynamic arena for further developments as European explorers scoured the globe for discoveries that revealed a theretofore unprecedented plethora of plant and animal diversity. The classification of organisms God placed on the planet was flowering through the Swedish taxonomist Carl Linnaeus's work organizing life into formal genera and species based on morphological traits and their similarity and differences among separate groups. In addition, French natural philosopher and anatomist Georges Cuvier explored the relatedness of species by examining in detail their skeletal structure, claiming that if someone brought him a single bone from an animal, he could reconstruct the whole creature. As merchant ships brought in fossil bones unearthed around the world, Cuvier's examination of newly discovered specimens provided evidence that many species had previously gone extinct, opposing the doctrinal notion of a plenum of species.

By the late eighteenth century, evolution was in the air, and even Darwin's grandfather Erasmus Darwin was writing about it in ways that his grandson would expand on. For example, E. Darwin speculated that all of life was derived from a single organism: "It is not impossible, but that the great variety of species of animals, which now tenant the earth, may have had their origin from the mixture of a few natural orders."[22]

However, while notions of species change emerged early in antiquity, no formal mechanism was offered for this change. Darwin and Alfred Russel Wallace, who independently discovered the same principle, would provide that mechanism: natural selection.

Charles Darwin

As a naturalist, Darwin was uniquely qualified to frame the questions that would become essential to his book *On the Origin of Species*, which provided the first coherent explanation of the mechanisms by which species evolve and change into new species. While on his five-year voyage on the HMS *Beagle* under the command of Captain Robert Fitzroy (December 27, 1831–October 2, 1836), Darwin had the chance to visit several of the world's diverse ecologies and habitats. During his time sailing with the *Beagle*, he observed much that would inform his thinking on species change. For example, upon his return to England, Darwin noted how, among specimens of finches he and others collected while the *Beagle* visited the Galapagos Islands, this geographically unique cluster of bird species was found to be related to a species of finches from Ecuador, the closest mainland to the island chain. Darwin noticed that the world's island flora and fauna often seemed to have branched off from their nearest kin on nearby continental or other large island sources. He inferred that the Galapagos finches were descendants of a lost pair of Ecuadorian finches that came to the island eons ago and subsequently diverged from their original type. Clearly, they had speciated from their founding lineage of continental birds. But how?

Darwin compared the changes among species in the natural world with how humans modified agricultural plants and animals for transportation, traction, food, fiber, and even those avian messengers, homing pigeons, used to send messages across battle lines during wars. Through such human manipulations of wild species of plants and animals, things like grain and seed sizes increased. Dogs and horses had been bred to produce multiple forms useful for supporting the needs of people. Repeatedly, humans restricted and enhanced traits that helped human flourishing by purposely selecting advantageous or desirable traits and suppressing broodstock or crop propagules that failed in these aims.

As humans selected the traits and attributes they needed or found useful, domestic varieties of crops and animals diverged from ancestral forms. The key was active selection. However, who did the "selecting" if humans were not present in the natural world? The variation in the characteristics of our agricultural crops and livestock seemed to need the guiding hand of humankind. Nevertheless, variation was also found ubiquitously in the natural world. What would selection for valuable traits in such a system look like?

To study this variation, Darwin spent ten years dissecting barnacles. It was clear that even within a single species, anatomical features varied. If some agent selected among these variants, how many varieties of barnacles might it be possible to generate? Darwin's great insight was that nature selected those traits and attributes most useful in the environment in which the organism had to live and survive. Darwin notes the reproductive potential of organisms by looking at the slowly reproducing elephant:

> There is no exception to the rule that every organic being naturally increases at so high a rate that if not destroyed, the Earth would soon be covered by the progeny of a single pair. . . . The elephant is reckoned to be the slowest breeder of all known animals. I have taken some pains to estimate its probable minimum rate of natural increase: it will be under the mark to assume that it breeds when thirty years old and goes on breeding till ninety years old, bringing forth three pairs of young in this interval; if this be so, at the end of the fifth century there would be alive fifteen million elephants, descended from the first pair.[23]

The fact that biological organisms have more offspring than can survive creates a struggle for existence among those individuals who carry beneficial heritable traits that allow them to survive better than others in the population. These successful individuals outcompete their neighbors for valuable resources because those traits are beneficial in a particular environment and will then pass on the beneficial characteristics to their offspring. Natural selection acts as a filter for the heritable variations that turn out to be most valuable to an organism's continued survival.

Notice that in addition to selection, evolution also depends on this variation. While natural selection pulls various traits into the next generation that are advantageous in a particular environment, standing variation is essential for producing novelty. It is the basis on which natural selection acts. Many years later, it would be discovered that mutational processes maintain variation. DNA is a long chain of nucleic acids of such a size that minor quantum effects play a role in flipping one kind of nucleic acid to another, producing the novelty that drives evolution through changes in complex gene networks.

Since Darwin's original articulation of evolution through natural selection, many refinements, discoveries, and analyses have confirmed the vitality of Darwin's original idea. Science often aims at elucidating two targets: explanation and prediction. An explanation comes by understanding how underlying causal mechanisms contribute to the patterns and processes we find in nature. Prediction is achieved by understanding the inferences and implications of theories constructed to explain general patterns that bear on the temporal course processes take through time and space and projecting them into the future. The science of evolution has been wildly successful at both aims. During the nearly two centuries since the publication of *Origin of Species*, the science and exploration of the concept have significantly confirmed and added to the evidence for the original theory. Our understanding of micro and macro processes involved in evolution has grown exponentially.

One of the most significant leaps forward in explanation and prediction came with what has come to be known as the "modern synthesis." Darwin and his contemporaries did not have a suitable explanation of inheritance in organisms, which was critical for the theory of evolution through natural selection. The most popular theory about heritability in Darwin's contemporaneous ether was that inheritance followed a continuous blending of parental traits. However, it was demonstrated mathematically that if a blending of traits occurred, it would eventually drive evolution to a halt by decreasing the variation of traits, as selection moved the population to the average expression of these traits—the variation necessary for evolution to occur would evaporate in the averaging.

Darwin and Wallace's idea of evolution through natural selection appeared doomed. However, in a sadly ignored paper published in 1866 titled, "Experiments in Plant Hybridization," an Austrian monk named Gregor Mendel demonstrated with peas that inheritance was particulate and that the suspected problem of the pure blending of traits destroying variation did not always occur. The independent rediscovery of Mendel's paper by Hugo de Vries and Carl Correns launched a reevaluation of Darwin's work, showing that the fitness of Darwin's ideas thrived spectacularly under Mendelian genetics. Discoveries and advances in genetics, embryology, and mathematical treatments of the theory established that evolution by natural selection was possible and that it explained enormous swaths of current research. Darwin's ideas flourished as the lead explanation for species change. In a 1971 paper, leading evolutionary geneticist Theodosius Dobzhansky rightly said, "Nothing in biology makes sense except in the light of evolution."[24]

Further confirmation and expansion[25] of evolution has come through multiple lines of productive and nuanced disciplinary evidence: genetics and the discovery of the mechanism of inheritance in DNA, RNA, and protein synthesis; the role of quantum field theory in mutational effects of nucleic acid changes; how genes and proteins work together in emergent networks of genetic effect to give us the complex developmental and embryonic mechanisms that control an individual's developmental process; paleontological evidence from the fossil and archaeological record, including an abundance of fossil finds from human lineages; applications in many diverse fields, from medicine and agriculture to engineering and even contemporary artificial intelligence applications of evolutionary-based genetic algorithms used to solve practical technical problems and find solutions to complex design and manufacturing problems. Thousands of peer-reviewed papers are published annually in academic journals, exploring, confirming, enhancing, and creating a more precise picture of how evolution informs our understanding of life and its processes.

One nuance is worth mentioning because it appears repeatedly in modern creationist attacks on evolution. Biologically, there is no equivalent to the ancient Hebrew notion of "kind." Species today are defined scientifically by evolutionary relatedness as measured by the similarity of DNA sequences. Rather than seeing species as a fixed biological notion, biologists see them as entities embedded in the complex processes that constitute them. Species are maintained, in part, by imperfect reproductive isolating mechanisms so that organisms tend probabilistically to mate with members of the same species. However, these isolating mechanisms can be subverted. Nothing keeps DNA, the chemical algorithm that guides the development of the final biological form, from wandering away from a parental form in the next generation and, through repeated selectional bouts, eventually creating new species. As a species finds itself in a new environment, it must respond to new survival challenges by drawing on the standing random genetic variation found in its DNA. The concept of the immutability of species is a carryover from the instance that "kinds" are maintained by a law of nature. This is demonstrably false. The creationist insistence that species are fixed has no place in modern biology and has been rigorously refuted. For example, species can change when a small population gets isolated enough to stop gene flow with its former population. Then, when the isolated population's DNA wanders too far from that original population, it eventually cannot mate with the source population. Should the isolated population reconnect with the original population, their ability to reproduce has been compromised, and now two species exist where there was only one before. Throughout Earth's history, changing environments, species movements among those environments, and other complex forces have allowed speciation to be a common event over the time lengths involved.

We will see in what follows how Jeffery worries that fundamentalist Protestant perspectives have crept inappropriately into the Smithian restoration, diluting the original theological ambitions of the church's founder, Joseph Smith Jr. This is regrettable, as evolution fits very nicely into the restoration. For example, both Mormonism and evolution are invested in progression and embodiment.[26]

Evolution has been radically productive as a scientific theory, so much so that creationist attacks are often forced to scorn the entire scientific enterprise. To dismiss evolution as a way of knowing the material world—currently the only viable contender as a coherent way to understand the history of life's biological development on our planet—is a mistake that lures people away from all science. Its dismissal impeded the LDS Church's relationship to science. As we will see as we continue to look at evolution within the LDS Church, anti-science stances are often the natural consequence of dismissing organic evolution.

This necessarily brief flyover of the history of science has just skimmed over the surface of an ocean of dynamic research findings in one of the most productive and important sciences going on today. We now turn to the relationship of the church and its principal drivers in its relationship to evolution in its scientific, social, and religious dimensions.

A Brief Sketch of the Relationship Between Evolution and the Church

At the end of the nineteenth century, the Utah branch of Mormonism, like many people living at the time, had largely positive feelings about science, though the concept was poorly understood and often conflated and used synonymously with "knowledge."[27] However, science was often touted as an overall good. Brigham Young maintained that science was what distinguished the LDS Church from the rest of the Christian world, as illustrated by his address at the Tabernacle in Salt Lake City on May 14, 1871:

> In these respects, we differ from the Christian world, for our religion will not clash with or contradict the facts of science in any particular. You may take geology, for instance, and it is a true science; not that I would say for a moment that all the conclusions and deductions of its professors are true, but its leading principles are facts—they are eternal.[28]

One wonders what Young saw in geology to hold it up as an example of "true science," but twenty-five years later, in 1896, James Talmage would become the first Mormon to achieve a doctorate from Illinois Wesleyan University in geology. Talmage would hold several influential positions, including becoming the president of the University of Utah. He worked to promote the sciences and teach the saints about them, including writing two books for a popular audience: *First Book of Nature*[29] and *Domestic Science: A Book for Use in Schools and for General Reading.*[30]

According to Tim S. Reid's doctoral dissertation, "Mormons and Evolution," the first mention of evolution as a theory by a church leader was made by George Q. Cannon in 1861, just three years after the publication of Darwin's *Origin of Species* in which he suggests that lower species can have evolved but denied that evolution was a part of the special creation of humankind.[31] Cannon would eventually reverse course and claim that evolution was inimical to gospel teaching on the origin of man and that the idea that evolution explained species diversity was "ridiculous." He would be joined by Apostle Orson Whitney, who argued that based on scriptural sources, species were fixed and immutable.[32] They reinforced passionate arguments by an LDS teacher and writer, Joseph B. Keeler, who joined Protestant concerns with evolution in attacking the discipline with vigor, cherry-picking scientific sources to argue that evolution was wrong both scientifically and scripturally.

The Rise of Protestant Fundamentalism

Debates over evolution and religion emerged nationwide as the Protestant fundamentalist movement argued for biblical inerrancy and scriptural literalism. The doctrine of inerrancy maintained that the Bible contained "autographs" given to the original authors of the work and contained no factual errors: "They are factually accurate in all that they have to say, including when they speak on matters of history, science, and the like."[33]

The movement also embraced an apocalyptic eschatology that emphasized biblical references and prophecies interpreted through a fundamentalist lens about the violent cleansing of the world in the last days of Earth prior to the imminent second coming of Christ. There was a general fear of liberal biblical criticism and the risk that it would undermine faith through secular readings of scripture. Moreover, Darwin's *On the Origin of Species* explained our planet's biotic diversity and novelty, which had previously been used as evidence of God's direct involvement in the Creation.

Between 1910 and 1915, Texas billionaire Lyman Stewart published a major twelve-volume work called *The Fundamentals*. The book contained articles that tackled fundamentalism, apologetic discourse, the nature of the Trinity, and other issues of Christian concern.[34]

There are two surprises in this work of interest to us as we move forward. First, one of the "isms" addressed in this work was a negative perception of Mormonism, which may make fundamentalism's embrace among certain members of the church unexpected. Moreover, one of the essays held that theistic Darwinism might be possible.[35] This contradicts many extant forms of Protestant Fundamentalism today that hold to a "young earth" creationism, asserting the planet was created about six thousand years ago.

Another complication in the emerging debate about evolution and faith concerned the opposition of established scientists to Darwin's insight about natural selection as a productive driver of speciation, which seemed incompatible with late-nineteenth-century views on heredity. These pre-Mendal complaints cast doubt on Darwin's explanation of natural selection as the mechanism of evolutionary change providing fodder for further complicating debates about faith and evolution.

In the early twentieth century, science was in the air, and several members of the church contributed publications championing scientific perspectives and demonstrating compatibility with LDS thinking and doctrines. Indeed, several prominent Mormon thinkers drew on the prestige of science to show that evolution was compatible with the church. For example, in 1904, Nels L. Nelson (b. 1862), a graduate and then instructor at Brigham Young Academy,[36] published *Scientific Aspects of Mormonism* with G. P. Putnam.[37] He aimed to present the case that Mormonism comports with "modern scientific thought"[38] and squares with "that newer revelation of God, 'the Book of Nature,'"[39] that is, the universe as revealed through science.

This was also seen in the future apostle John A. Widtsoe's classic, *Joseph Smith as Scientist: A Contribution to Mormon Philosophy*,[40] in which he attempts to legitimize the thought of the founding prophet by claiming he could be shown to have anticipated much of contemporaneous scientific thought. Widtsoe earned a PhD in Germany from the University of Göttingen and would figure in some of the debates about evolution. For example, one of the more intriguing things in the book is the chapter called "The Law of Evolution" in which he used the controversial evolutionary biologists Herbert Spencer and Thomas Huxley as examples of thinkers whose theories were foreshadowed by Joseph Smith's thought. Spencer, a fervent, atheistic defender of evolution, was one of the principal authors of social Darwinism (he coined the term "survival of the fittest"). Huxley, an evolutionary biologist who embraced Darwinism enthusiastically, became one of Darwin's most aggressive defenders, earning him the name "Darwin's Bulldog." Widtsoe primarily embraced evolution not only in its biological sense but also in the sense of Nelson's 1904 book, as the cosmic evolution from humans to gods. His use of Spencer and Huxley suggests that he viewed evolution as less controversial than it would shortly become.[41]

However, this initial fascination with science and evolution was fraught, and debates about evolution and religion were fomenting. Interest in evolution among members of the church was giving rise to questions about its place in Mormonism, and in 1909, Joseph F. Smith became troubled by the ideas coming out of the science of evolution, as well as new developments in biblical criticism. He asked Orson F. Whitney to draft something that could be used to frame the position of the church on these issues. The statement was signed by the three members of the First Presidency, then consisting of Joseph F. Smith, John R. Winder, and Anthon H. Lund. They issued a statement on the "Origin of Man," published in the November issue of the church magazine, *The Improvement Era*.[42] It begins,

> Inquiries arise from time to time respecting the attitude of the Church of Jesus Christ of Latter-day Saints upon questions which, though not vital from a doctrinal standpoint, are closely connected with the fundamental principles of salvation. The latest inquiry of this kind that has reached us is in relation to the origin of man. It is believed that a statement of the position held by the Church upon this subject will be timely and productive of good.

This suggests that evolution was enough of an issue in public discourse that the First Presidency felt it needed to be addressed. The statement was largely anodyne but can be read as holding evolution somewhat in suspicion, while more clearly promoting LDS perspectives on God's role in the Creation and the divine nature of humankind, ending with,

> Man is the child of God, formed in the divine image and endowed with divine attributes, and even as the infant son of an earthly father and mother, is capable in due time of becoming a man. Hence, the undeveloped offspring of celestial parentage is capable, by experience through ages and eons, of evolving into a God.

Provocatively, this suggests evolution at an expanded "temporal scale," adding deification as an element of possible human evolution.

Stephen Taysom's biography nicely summarizes some context for President Joseph F. Smith's views that informed this statement and that sets the stage for the evolution controversy that would come to haunt BYU's relationship to evolution. Taysom says concerning this 1909 statement, "By this point, JFS [Joseph F. Smith] was expressing in private his belief that there was an unbridgeable gap between 'The doctrine of Christ and fictitious science.'"[43]

Taysom also points out that Joseph F. Smith had unique views of doctrinal development as early as the late 1870s and 1880s that also came to influence how scriptures related to Creation are read. He writes,

> The key to JFS's doctrinal thinking was always vested in the ability of the church president to control scriptural interpretation in such a way as to emphasize stability by insisting on the obviousness and unchangeability of doctrine, especially during periods when scriptural interpretations were changing the most. So, while he claimed to welcome all truth with joy, if it contradicted the traditional, literalist reading of scripture, then it was not truth. Such an epistemological model was required to maintain the internal consistency and doctrinal order that JFS sought to create within Mormonism, and it was this pressing need for order and coherence, in other words, sociological conservatism rather than theological conservatism, that led JFS to guide the Church down a deeply conservative channel both theologically and sociologically.[44]

JFS's son, Joseph Fielding Smith, who also harbored these suspicions, would enable a lot of Protestant Fundamentalist opinions about the science of evolution and the age of the Earth to leak into church discourse and some of the instruction manuals used in the Church Education System (CES). Bowman suggests that to label him simply as a fundamentalist, however, misses that he disagreed with Protestant Fundamentalism because Smith believed that without the added resources of direct prophetic revelation, their interpretation of scripture is flawed and can be ignored. With modern revelation, however, the plain reading of scripture was unmasked, and no interpretive apparatuses were needed.[45]

The issue of who interprets scripture and how they go about it played an essential role in how evolution was received by many prominent leaders in the church, as will become evident in what follows.

The 1911 BYU Controversy

A growing controversy about evolution and progressive religious interpretations of biblical criticism within the church reflected a microcosm of the national discussion. It rent BYU to its core, nearly destroying its credibility just as the school was emerging as a respected modern university. The episode would pit professors trained in modern universities, where Latter-day Saints flocked in the early part of the century, against those trying to keep modern ideas at bay. More conservative and less academically well-educated church administrators were tempted by the growing influence of the fundamentalist movement into discounting evolution. This trend would eventually take on the more creationist-infused application of "fundamentalism," the word coined in 1920 that would eventually deform into young-earth creationism and its offspring the intelligent design creationism that we see today.

The 1911 controversy can be summarized as follows. The first president of BYU, George Brimhall, was determined to fulfill his vision of BYU becoming a world-class institution by recruiting the best academically trained LDS scholars to serve on the faculty. Thus, he hoped to create a thriving university that could compete with the finest universities in the world.

In 1903, he hired BYU's first formally trained academics. First, he chose Joseph Peterson, a PhD graduate of the University of Chicago, to build the psychology department. To take over the College of Education, Brimhall then hired Joseph's brother Henry, who had obtained an AM from Harvard.[46] In 1908, Brimhall hired the University of Utah's biology department chair, Ralph V. Chamberlin, with his recently minted PhD from Cornell, to head their own biology department. Ralph's brother, William Henry Chamberlin, trained in ancient languages and theology at Harvard, then joined the Peterson brothers as members of the part-time theology faculty. Following these hires, a new dynamism emerged at BYU, with students actively engaged in issues and ideas exploding in academic circles at most major universities. As Bergara and Priddis note, "subjects from communism to eugenics were hotly debated inside and outside of class."[47] At the 1909 Darwin Days celebration, Ralph Chamberlin gave a passionate speech about Darwin's contribution, calling him "one of the greatest scientific minds of the era."[48]

However, this turn to more progressive academic discussions also created an aura of suspicion around the four men credentialed by first-rate institutions of higher learning. In part, these teachers were not just Darwinian evolutionists but also what might be called "Huxleyan" social theorists, who were arguing for the use of evolution, not just as a scientific finding but also as a moral "ought." While "eugenics" was coined in 1883 by Darwin's cousin, Francis Galton, its degrading harm in using radical legalized racist practices to maintain racial purity would not be fully expressed until Julian Huxley's work in the 1930s and '40s, promoting the social engineering of humanity, which would find one of its most evil expressions in the rise of Nazism.

Horace Hall Cummings, the commissioner of Church Schools, was called on to investigate whispers that something fishy was happening with these progressive professors. Although not responding to eugenic concerns, the rumors suggested that the new professors were corrupting the youth with ideas from modern biblical criticism, liberal theology, and evolution. Cummings took these reports seriously in part because he harbored doubts about the usefulness of modern education. It would not be unfair to call him anti-intellectual. He had failed earlier attempts to enter higher education but eventually obtained a bachelor's degree from the University of Utah in 1906, just before being made the commissioner of Church Schools. Bergera and Priddis note that Cummings viewed Brimhall's new hires with alarm. He feared the academic thought they were teaching would move the students away from the "religious simplicity" and "gospel purity" he was trying to protect.[49]

Based on no evidence other than hearsay, Cunningham came to BYU for nine days to attend the professors' classes. He was received amicably at first, but things changed as it became clear that his agenda was to remove evolution and higher biblical criticism and those teaching these subjects. The animosity became palpable as Cunningham came under fire from undergraduates and faculty alike for being anti-intellectual and promoting unquestioning piety rather than an informed yet faithful approach to modern scholarship in evolution and biblical studies. In a meeting with the faculty, when asked what he would report to the board of trustees, he disingenuously replied that "he was glad to learn through conversation with the [university] presidency that matters [had] been misrepresented."[50]

However, such reassurance stands in stark contrast to the report he actually made to the board in which he stated, "When some of the most radical changes in theological views were introduced, it caused a great disturbance in the minds of both pupils and the old-style teachers."

Like some church members obsessed with deference to inferred leadership priorities, he had no interest in establishing facts on the ground. Cunningham's misunderstandings and outright fabrications resulted in an ultimatum: The Peterson brothers were asked to either stop teaching modern topics or to leave. Brimhall tried to persuade Henry Peterson to stay but to cease teaching the controversial subjects. As Peterson said, "All I needed to do was to be a good boy, teach the permitted doctrine only, and I could stay with them. Think of it! I was invited to stay as a hypocrite teaching one thing to my students and believing and feeling another!"[51] As Henry and Joseph refused to teach under these conditions, they were dismissed. Ralph Chamberlin left in solidarity.

One often neglected part of this tale is the students' reaction to these firings. The school newspaper, the *White and Blue*, ran a beautiful student-written essay signed by nearly 85 percent of the enrolled student body expressing their desire to learn the best academic ideas currently taught in universities worldwide. The essay writers argued that students were not threatened by controversial ideas:

> Because the best modern educational thought takes the theory of evolution as a basis, we feel that it should be taught here. This does not mean we thereby assume the theory is true or false, but simply that because it commands the greatest thinkers' attention, it should be open to investigation.
>
> As college men and women, we have confidence that if the evidence which tends to support the theory of evolution be presented simply for what they are worth, we will have sufficient discretion to determine whether or not we wish to accept them. In so far as we have studied the subjects in question, we feel that we have broadened in that we have seen both sides of a mooted question. We believe that it is not the proper attitude to fight a proposition by ruling it completely out of consideration. We feel that if our gospel is true, it will triumph over error without any artificial protection.[52]

It was a striking call by the student body to give them the hard facts as taught by the best universities in the world and let them sort it out rather than being policed by well-meaning but academically ill-equipped university leaders who wanted to protect the students from the "world." Jeffery would embody the spirit of these students when he arrived nearly sixty years later.

William Chamberlin was not asked to leave. He published two papers in support of the three dismissed professors. The first appeared in the school's paper, the *White and Blue*, on February 14, 1911. "The Theory of Evolution as an Aid to Faith in God and Belief in the Resurrection" argued that evolution is an imminent (as opposed to a transcendent) purpose in nature. About a month later, on March 10, 1911, he published an article in the *Deseret News* supporting evolution. One of the first academic theologians to teach at a Church School, he held on at BYU, teaching its most popular classes in psychology, philosophy, and theology. However, he continued to be treated with suspicion, and for the 1913–14 school year, he was taken off the theology faculty at BYU and was not allowed to teach a class on the life of Christ.

On March 12, 1916, he wrote his brother Ralph that most of his courses had been cut despite their popularity: "Twenty-two of the leading college students are taking Philosophy of Nature; over fifty are taking Social Psychology; I have eight graduate students and some seniors taking a seminar in Bergson. . . . On account of such things I am writing my resignation this evening."[53] With his favorite and most interesting classes cut, he felt like the harassment and dismissal of his efforts were too much, and he resigned from BYU, thereby damaging the students' opportunity to get a first-class education with a caring professor. A passage in William Chamberlin's biography, written by his brother Ralph, captures nicely what the students of BYU meant to William:

> W. H. Chamberlin devoted himself to aiding students to develop in the light of modern knowledge the best principles of the past into higher and greater principles for the future. He did not indoctrinate them; he led them to understand religious experiences and truths and organize these for their own lives. The appeal was always to the inner lives of the students. Under his guidance, they were able to find for themselves ways to higher levels of thought, expression, and action. With increased loyalty to cherished convictions and faiths, they saw them in new lights, charged with new meanings.[54]

William left for a year to study at Harvard and returned to Utah to find employment. However, the old animosities remained within Church Education; Horace Hall Cummings was still the commissioner of Church Schools (working from 1905 to 1920) and enforced his vindictive anti-intellectual buffoonery by blocking William from teaching at Dixie College in St. George, Utah, which was then a Church School. Recently married, William took a job with Utah Extension, which, while not ideal, provided an income. Finally, after Cummings was replaced by David O. McKay as the new commissioner, William Chamberlin was asked to return to BYU.

Unfortunately, William died shortly after being reinstated at BYU, having contracted pneumonia during an influenza attack. Looking over Chamberlin's life, David O. McKay wrote this in a letter to William's brother Ralph on February 17, 1926:

> That a lofty, sincere soul like W. H. Chamberlin should have been compelled to struggle in our community and to have been misunderstood by those who should have known him best, seems to me to be nothing short of a tragedy. . . . I wish it had been my privilege to know him intimately. For one thing, however, I am thankful, namely, that I had no reservation in mind when it came to my privilege to recommend that W. H. Chamberlin's services be again secured for the Church Schools.[55]

These events had a chilling effect on science education at BYU for the next sixty years, doing damage to its reputation and mission as an institution of higher learning.

The B. H. Roberts, James Talmage, and Fielding Smith Evolution Affair

The second significant moment in the history of evolutionary theory and the church pitted some of the greatest thinkers in the LDS movement against each other on the subject of science and evolution. Among them, B. H. Roberts and Fielding Smith played opposing roles in a growing debate about evolution in the church. Apostle James Talmage also entered the debate, attempting a compromise between the views of these two fiery men. Eventually, the First Presidency of the church was asked to weigh in on the questions mapped out by these three men on evolution but declined to enter the debate. This story has been told in many places, including Jeffery's landmark *Dialogue* essay "Seers, Savants, and Evolution," which we will examine more fully in the next chapter. Nevertheless, a brief sketch outlining the debate will give some context to Jeffery's arrival at BYU and the ensuing controversies.

To beef up the church's Sunday school manuals, B. H. Roberts wrote a series of essays to be included in a yearlong course of study, which he hoped to publish as *The Way, the Truth, and the Light: An Elementary Treatise on Theology.* Roberts considered this his magnum opus, even more so than his well-received six-volume work, *History of the Church.* Like the work of George Eliot's character Edward Casaubon in the novel *Middlemarch*, Roberts's work could be described as magisterial but flawed; it is magisterial in that, like Casaubon's great work on world religions, *The Key to All Mythologies*, Roberts intended his work to be a compendium of all current thinking relevant to Mormonism about everything related to the gospel, from the Creation to the Earth's final disposition. This effort included an attempt to reconcile the church's teachings on Adam and Eve with the findings of the archaeological and paleontological record showing that modern human remains stretch back for at least two hundred thousand years. Roberts speculated that pre-Adamite humans existed before Adam and Eve came into the world. Davis Bitton critiques Roberts's efforts as naive and needing revision, pointing out that he was an amateur whose information came primarily from popular magazines. The work missed important threads in some of the vibrant thought in science and philosophy being discussed in intellectual communities of the early twentieth century, adding that Roberts's view of science was inadequately grounded even for the science and philosophy of the 1920s.[56] However, his attempt to bring science into conversation with church doctrine was a sincere attempt to reconcile these two different ways of seeing the world.

On the other hand, holding the creationist leanings of his father, church president Joseph F. Smith, the younger Fielding Smith, saw Roberts's views as heresy. In Fielding Smith's view, there was no death before the Fall of Adam and Eve, so pre-Adamites were doctrinally impossible. His attempts to dismiss such ideas began a lifelong battle to keep the science of evolution at bay. Fielding Smith's views reflected not only his antipathy toward evolution or science but also included his complex reaction against modernity.

Historian Matthew Bowman sees Fielding Smith's views as part of an informed anti-progressive stance that spanned several disciplines, manifesting opinions ranging from how scripture should be read with a nod to embrace simplicity and prophetic infallibility to his refusal to give any credence to progressive biblical interpretations from the leading academic institutions engaging in the "higher criticism." Fielding Smith taught that from the time of Adam, the gospel had been taught in its fullness, and any departure from that would entail a departure from the scriptures. He viewed humanity as not only God's greatest creation but also God's literal spirit offspring. From the beginning, Adam and Eve and their descendants were created at the height of their moral and intellectual powers. He believed that racial differences pointed to intellectual and moral differences that God had set in place based on preexistent behavior and merit. For example, based on his deep anti-modernism, Fielding Smith held that humans would never land a human on the Moon because God placed people on Earth, and its bounds would be the limit of their allowed progression.[57] In this worldview, evolution was a monstrous doctrine that saw humans reaching, like the builders of the Tower of Babel, toward endeavors beyond the boundaries God had set.

It is essential to keep in mind that Fielding Smith was not just reacting to the science of evolution, which I am focusing on here, but was concerned with larger social currents, often associated with thought that challenged several key issues Fielding Smith, and others like him, held dear about the way history had unfolded since the time of Adam. For example, Bowman points out the effect on early LDS of German scholar Julius Wellhousen's ideas that the Israelites progressed from "primitive animism to what he called ethical monotheism." His observations supported the idea of a cultural *evolution* from a simple society to a more complex one. This take on history stood in contrast to the view of history Fielding Smith held—namely, humanity had started with complete knowledge of truth and lost these insights through time. It did not help the situation that Ralph Chamberlin, one of the dismissed professors, gave a speech drawing on Wellhousen's work or that B. H. Roberts had also fully embraced the German's thought. Fielding Smith was concerned that these beliefs about the unfolding of religious history had no place in his view of the restored gospel's conception of what had happened.[58]

In the early part of the twentieth century, Smith's attacks were aimed primarily at Roberts and Widtsoe. These two general authorities were comfortable with at least a nod to evolutionary explanations of the Creation, while still holding on to some suspicions that humans may yet have been a special creation, unlike the remainder of life on Earth. Likewise, Apostle James Talmage, a trained geologist and the church's first PhD in any field, was comfortable with an ancient age for the Earth and felt that there were bounds to the accommodation evolution demanded of church creation doctrine. At the same time, while Talmage was willing to grant that species or "kinds" were likely immutable, he especially resisted the idea that the development of man from lower forms was possible.

Furthermore, Joseph Fielding Smith embraced a particular flavor of Christian young-earth creationist ideas promoted by Seventh-day Adventist writer George McCready Price and that differed from mainstream forms of fundamentalist creationism which allowed for a much older Earth, rather favoring a much younger creation. Price, in pondering the geology textbooks of the day, determined that science's interpretation was wrong and that science itself had gone astray in departing from a literal interpretation of the flood and the creation of the Earth.[59] Thus, Fielding Smith embraced the idea that Adam lived about six thousand years ago, that there was no death before the Fall of Adam, and that evolution and the Gospel of Jesus Christ were irrevocably incompatible.

Interestingly, the disagreements between these highly ranking General Authorities were not hidden or massaged to create an impression of unity within the quorums. On the contrary, they were very public about their views, sharing talks and writing articles highlighting their disagreement—something rarely seen today in the church. Since the establishment in the late 1960s of Correlation (a movement to unify and simplify the church's international organization that had grown over the previous decades), an image of unity of purpose among the General Authorities has been carefully maintained.

In 1930, Fielding Smith gave a speech attacking the scientific record, which pointed to a far more ancient planet. Roberts was incensed and wrote a letter to Rudger Clawson, president of the Council of the Twelve Apostles. In his letter of complaint, he stated, among other things, "I call in question the accuracy of Elder Smith's position in reference to the whole doctrine of his discourse, as being contrary to a great volume of well-developed and ascertained truth, established by the research of scientists of highest character, of profoundest learning, and worldwide research."[60] He further cautioned that doctrinal interpretations of scripture contrary to science were injurious to their faith and reduced the restoration "to the character of a narrow, bigoted sect, forsaking the God-given world movement idea of it."

On January 14, 1931, Smith responded in a paper read to the council, "Any doctrine, whether it comes in the name of religion, science, philosophy, or whatever it may be, if it is in conflict with the revelations of the Lord will fail. You will find that every doctrine, theory, principle no matter how great it may appear, no matter how universally, it may be believed, if it is not in accord with the word of the Lord it will Perish."[61]

Fielding Smith suggested that evolution was part of a satanic attempt to undermine God's work: "The doctrine of organic evolution which pervades the modern-day sciences proclaiming the edict that man has evolved from the lower forms of life . . . is as false as their author who lives in hell."[62]

Smith and Roberts continued championing their views, and the First Presidency was asked to weigh in. Of special concern was Roberts's use of pre-Adamites, and he was encouraged to cut those parts from his book. Roberts refused, and the First Presidency issued the following in 1931, suggesting that the church was not interested in defending or defeating evolution:

> The statement made by Elder Smith that the existence of pre-adamites is not a doctrine of the Church is true. It is just as true as the statement "there were not pre-adamites upon the earth" is not a doctrine of the Church. Neither side of the controversy has been accepted as a doctrine at all.[63]

They added,

> Upon the fundamental doctrines of the Church, we are all agreed. Our mission is to bear the message of the restored gospel to the people of the world. Leave Geology, Biology, Archaeology and Anthropology, no one of which has to do with the salvation of the souls of mankind, to scientific research, while we magnify our calling in the realm of the Church.
>
> We can see no advantage to be gained by a continuation of the discussion to which reference is here made, but on the contrary are certain that it would lead to confusion, division and misunderstanding if carried further. Upon one thing we should all be able to agree, namely, that Presidents Joseph F. Smith, John R. Winder and Anthon H. Lund were right when they said: "Adam is the primal parent of our race."[64]

The first presidency hoped this would set the matter to rest. However, it did not address the fundamental conceptual issues in the conflict between Roberts and Fielding Smith on evolution, which centered on, among other things, differences in how revelations were received. To Fielding Smith, revelation was simple, easily interpreted, and accepted as straightforward, literal declarations of universal truths. Roberts saw science as the best approach to understanding the universe and believed science and faith revealed truths that might appear to conflict but could be reconciled as more was learned from revelation and science in conversation.

Talmage had not been involved in this debate, but he was bothered as he read what Fielding Smith had written. He sent the talk to his son Sterling, a working geologist. The younger Talmage pointed out that most of Fielding

Smith's sources were from Price. The younger Talmage dismissed Price's *The New Geology* as nonsense. Despite the book's lavish illustrations and photography, Price was not a geologist, and it showed. Talmage reported to his father that Price had no standing in the scientific community and should not be taken seriously by anyone.

Talmage reopened the debate by giving a talk called "The Earth and Man" in the Tabernacle on August 9, 1931, which was reprinted in the *Deseret News* on November 21, 1931. In this, Talmage framed a moderate response that suggested that the history of the Earth showed a trend from simplicity to complexity, that the fossil record showed clearly that the Earth was much older than man, and that traces of evolution were abundant. However, he argued that Adam and Eve were historical figures, and humans were unique creations that did not evolve. He offers this view by exploring humans' unique features that no ape possesses. Therefore, humans could not have evolved from lower beings.[65]

While keeping man's evolution at a distance, the speech was more moderate than Fielding Smith's and opened the door in people's minds to the possibility that some forms of life could have evolved. Talmage supported publishing Roberts's book. However, circumstances intervened: Talmage died two years later, followed by Roberts himself two months after that, and Widtsoe was dead by 1952. The principal supporters of Roberts's work were gone. It would not be published for more than fifty years when BYU Studies Press released it as a critical edition in 1995.[66]

At last, unopposed in his opinion on evolution, Fielding Smith published his polemical and anti-scientific book *Man, His Origin and Destiny* in 1954. Subsequently, this book would have tremendous anti-scientific influence on Church Education for decades to come.

Man, His Origin and Destiny

Man, His Origin and Destiny is a marvelously constructed response and critique of many of the topics of concern to Fielding Smith's disenchantment with modernism, including Protestant Fundamentalism, biblical inerrancy, modern biblical criticism, young earth creationism, organic evolution, and other forms discourse orbiting doctrinal matters.[67] Its tone is authoritative and erudite. It is well documented and referenced with selective readings from current thinkers on issues related to LDS doctrine, the plan of salvation, and humankind's place in the Creation. It has twenty-eight chapters and an introduction by Fielding Smith, a foreword by Mark E. Petersen, and another introduction by Melvin A. Cook. From a scientific perspective, of most concern to this work are the two chapters "Conflict Between Science and Religion, <parts:> 1–2," the five chapters "The Hypothesis of Organic Evolution, <parts:> 1–5," and the chapter "The Flood." Other chapters like "Adam's Place in the Earth's Destiny, <parts:> 1–3" and "Man the Offspring of God" and several others complain about Darwin's ubiquity in educational settings. Indeed, there are few chapters where evolution is not straightforwardly attacked suggesting the subject was the raison d'être for the work, even so it also expresses reasoned chagrin at other modernist concerns. Fielding Smith draws on several scientists[68] who have shifted from being Christians to being supporters of Darwinism to evince that science and religion are at war. He somewhat excuses their attacks on religion because he acknowledges that these men are "honorable and presumably honest in their convictions" and notes that their attacks are on "vastly different religion" than the correct version of ancient Christianity and its restoration contained within the church.

In much of the book, Fielding Smith creates a pastiche of creationist thinking drawn from several creationist thinkers. The most cited work in his magnum opus is from "Seventh-day Adventists' flood geology," Sterling Talmage argued above about Fielding Smith's response to Roberts's book.

In a review of Francis Gibbons's biography of Joseph Fielding Smith,[69] Gary Bergera describes how Fielding Smith's work writes off much of science as the product of dupes of Satan designed to destroy faith in Christ. Bergera also points out in this interview that "these qualities set an anti-intellectual tone in the Church that has lasted from the 1920s to the present."[70]

Fielding Smith's anti-modernist, anti-evolutionary stance is on full display in his book, where he repeatedly uses the trope that Satan was involved in evolution. For example, he says that to find the source for the theory of evolution, "we must go beyond the activities and research of mortal man to the author of evil [Satan], who has been an enemy of truth from the beginning before the Earth was formed."[71] This will also be reflected in his son-in-law, influential Apostle Bruce R. McConkie's book, *Mormon Doctrine*, in his entry on evolution.[72]

After *Man, His Origin and Destiny*, evolution was rarely discussed positively in official church publications before Jeffery arrived at BYU, but there are exceptions. One is worth noting: "The Relatedness of Living Things" by Bertrand Harrison, published in *The Instructor.*[73] The article creates a homey dialogue between a couple of neighbors, Brother Nielson, a professor, and Len Scott, a "dairyman extraordinary." Brother Nielson explains Darwin's thought to his neighbor, who is convinced by the professor's exposition and articulates a powerful sense that evolution is tame and can be accepted by both professors and dairymen. The current "Church History Topics Essay" on "Organic Evolution" on the official website of the church says, "In 1965, Church President David O. McKay worked with Bertrand F. Harrison, a botany professor at Brigham Young University, to foster greater understanding between Saints with differing viewpoints on evolution."[74] This is an oblique endorsement of evolution as a concept compatible with the gospel. That David O. McKay, then president of the church, supported this work is telling.

Fighting against creationism defined much of Jeffery's time at BYU. Creationist, anti-evolutionary forces had become entrenched in the church through the CES training and poorly researched sources, and its reliance on shabby fringe materials from dubious sources created conditions that required Jeffery to directly engage with those materials. This was important because CES was a powerful force in educating future generations about church doctrine and its purposes. However, what began as an organization to help the youth of the church reconcile the Mormon perspective with the wisdom of the world began to succumb to fundamentalist fears about evolution and even science itself. As Armand Mauss points out, "the CES now represents quite a different philosophy from that on which it was originally based during the twenties, when the church was still very much in the assimilationist mode."[75]

While there are several signposts of this switch in emphasis from promoting genuine educational efforts at learning to a more anti-intellectual stance, the shift may have begun at a CES Summer School where Apostle Harold B. Lee used Fielding Smith's book *Man, His Origin and Destiny* as the course of study.[76]

Mauss goes on to demonstrate the direction CES was taking. George S. Tanner, University of Idaho director of LDS Institute of Religion, recorded disapprovingly in his journal a 1962 staff summer school presentation in which Mark E. Petersen states,

> CES instructors are to teach the pure gospel and do nothing else; they are to instill in their students faith and testimony, for nothing else matters;

> they should not try to "steady the ark" by intellectualizing; there is no academic freedom in CES, and faculty who don't like that should simply resign; the LDS neither need nor want the learning of the world, and those who once went off to Protestant divinity schools to seek such learning were ruined by the experience.[77]

By drawing on writings of both the emerging Protestant Fundamentalists and young-earth creationists, *Man, His Origin and Destiny* cast a long shadow of misinformation over the science of evolution, influencing many members of the church who were making their way through school or were otherwise connected with the CES. T. Benjamin Spackman's excellent dissertation suggests that there are several reasons for the ascendancy of Smith's young-earth creationism promoted among the LDS: "1) the intellectual and emotional appeal of creationist writings resolving science/religion conflict in a way that confirmed traditional LDS readings of scripture, and 2) Smith's uncontested ecclesiastical authority in the public realm, backing 3) his claims to scriptural, scientific, and historical authority. This resulted in 4) Smith's claims being generally adopted among Latter-day Saints as the orthodox understanding of scripture and the Church's position on evolution." He also points to the populist appeal in pulling the interpretation of the scriptures from the hands of the experts and places it in the hands of the laypeople.[78] Additionally, Spackman adds that until Jeffery pushed back against this view, the near public silence from leading church authorities challenging the book left a sense that they supported the book. However, this embrace of Smith's creationist views interfered with the potentially positive LDS view of science that might have been possible had it not been embraced.

Four years after the first evolution undergraduate class in sixty years was offered at BYU, Jeffery published his landmark paper, "Seers, Savants, and Evolution: The Uncomfortable Interface" in *Dialogue: A Journal of Mormon Thought*. *Dialogue* was launched in 1966 at Stanford University under the guiding hand of Mormon intellectual Eugene England while Jeffery was at Berkeley.[79] Jeffery's paper, outlining the history of evolution vis-à-vis church authorities during the early twentieth century, covered ground not previously articulated publicly. The article would create an orogeny in the intellectual landscape that would structure and frame Jeffery's academic work for the rest of his life as a science educator.

CHAPTER THREE

Seers, Savants, and Deeper Conversations

Jeffery's Landmark Paper

In 1973, *Dialogue: A Journal of Mormon Thought* published a landmark article by Lester E. Bush Jr. (1942–2023) titled, "Mormonism's Negro Doctrine: An Historical Overview."[1] The essay changed our understanding of the history of the priesthood and temple restriction in the LDS Church for people of African descent, demonstrating how much folklore, bias, and prejudice had crept into our doctrine through ad hoc attempts by some church leaders to explain the reasons for its existence.

In similar fashion, Jeffery's paper, "Seers, Savants and Evolution: The Uncomfortable Interface," also published in *Dialogue* that same year, would change the academic landscape of LDS thought about science as it traced the background and history of evolution in the church.

The Genesis of Jeffery's Paper "Seers, Savants, and Evolution"

As Jeffery began teaching the first evolution class at BYU in sixty years, he had an inkling that trouble might lie ahead, especially from those among the religion faculty committed to the Protestant Fundamentalist creationist view assembled by Joseph Fielding Smith. What he found at BYU can only be described as a war. Science and religion departments were acrimoniously pitted against each other in demonstrably unproductive ways. Considering the conflict he found, Jeffery conceived of drafting something formal that would help the Saints understand that evolution was not a threat to their faith and that, more important, the church had no official stance on the subject—despite several vocal General Authorities and religion faculty who violently disagreed with that stance.

It would not be unfair to say he felt "called" to explore the topic, bringing his rigorous scientific training to bear. When Gary Bergera asked him in a 1981 interview, "Why did you decide to write 'Seers and Savants'?" Jeffery answered,

> Because it had to be done and nobody else was doing it. Indeed, it became evident that nobody else knew anything about doing it. That the people who should be were not. Our people in Church history whom I knew were aware of the critical documents were continuing to, in today's parlance, "watergate" those documents.[2]

He viewed much of the rhetoric against evolution, and by extension science, as harmful to the mission of the church and its basic doctrines. He began to sense that nothing would ever be written on this topic if he did not pick up the mantle to do so. He had gathered a prodigious amount of material about evolution and its relationship to the church while he was in Berkeley. Yet despite feeling confident in his position, he feared it would be an uphill climb, with criticism and indifference from both the science side, wherein his colleagues had little formal training in religion, and from the religion side, as the relevant faculty had very little interest in understanding science. Despite this, he mentioned in an interview that part of the reason he came to BYU was "to bridge the gap in the Church between science and religion." He had taught in the church's education program for college students, known as Institute, and he felt grounded in both the church's doctrines and beliefs, and with his deep education and training in the biological science at Berkeley, he was uniquely qualified to make the effort to work to form that bridge.

However, he also experienced some surprises upon arrival. First, few of the faculty were interested in bridging the disciplinary gaps between the sciences and religion as taught at BYU from either side. And second, he felt isolated, like an island between the two groups. He realized that if he didn't do the work, no one would.[3]

Two sources Jeffery used for "Seers and Savants" were essential to his project. The first was the 1931 memo from the First Presidency described previously in which they refused to take a hardline stance against B. H. Roberts's idea of pre-Adamites. Jeffery received these documents from Ken Godfrey, the institute director at Stanford University. The second was access to Apostle James Talmage's journals after they were deposited in the BYU Harold B. Lee Library. Jeffery discovered much valuable information about Talmage's perspectives on science within the journals, which helped him better understand the relationship between science and religion in the early twentieth-century church.

Despite all the busyness connected with his move to BYU, setting up his lab, and teaching students, Jeffery believed that the paper he was contemplating writing was important and that he was the only one ready and willing to do it. In a letter dated July 9, 1973, to his adviser, geneticist Curt Stern,[4] he said, "At the time we talked with you in April, I mentioned that I had just completed a paper reviewing responses in LDS theology to the matter of organic evolution. . . . You indicated at that time that you would have some interest in reading the manuscript, so I'm sending you a copy under a separate cover." He adds,

> I think I indicated to you at that time that in the last few decades particularly there has been a shift in basic philosophical sentiment within the ranks of Mormonism, away from a position that had a good deal of common ground with science (and a good deal of personal rapport with science), to a position that is much more reminiscent of the dominant theological Christian position of the 19th century. This, of course, has been quite distressing to many of the scientists in the Church, but no one has ever documented the matter or really tried to address the situation in any legitimate way. This paper, then, is aimed at a very broad audience, but essentially at non-scientists ("To the intelligent laymen," whoever they are).

He then gives a brief outline of the organization of the church's leadership, adding,

> The entire group of 15 men are considered to be prophets in the full biblical sense, and their words and teachings are taken by many church members as essentially those of deity. This applies in a particular way, of course, to the President himself, who is generally considered the mouthpiece of deity. It is against that type of background that this paper is written. The current President of the Church [Harold B. Lee] is a man not especially sensitive to avenues of knowledge other than religion, and I have tried to write this in such a way as should he and his colleagues read it, they hopefully can take no offense to it as propagandizing; I have tried to keep it as objective as possible.

He then complains that the church is becoming less friendly to science than it had been in the past.

This exchange between Jeffery and Stern demonstrates that he had no qualms about sharing his religious views with his non-LDS PhD adviser. Later, we shall see his faith attacked by detractors, including some among the church's leading General Authorities and many in the religion department at that time who considered him a heretic in need of silencing. In contrast to his comfort in sharing his religious commitments with scientists not of his faith, many from his own religious tradition were not a sympathetic audience for his views on science.

Suspicions—Jeffery Is Possessed by Satan

Even before Jeffery's landmark paper was published, things were heating up among certain faculty members, and the animosity was becoming a problem for the students. In 1972, a year after the evolution class at BYU began, BYU's new president, Dallin H. Oaks, recognized a growing distrust between the scientists on campus and the religion faculty. At the time, very few religion professors were trained graduates of academic religion programs at the university level; most came from the Church Education System (CES). President Oaks set up a seminar with science and religion department faculty to talk about organic evolution. It was contentious. Jeffery attended all the sessions, as did Roy Doxey, the dean of religion at the time, but the issue was so divisive that members of the group bickered extensively. Jeffery says in his 2009 interview that the contention between the scientists and some of the most vocal anti-evolutionary fundamentalist young-earth creationists in the religion department was fierce. Most of these drew on the work of Fielding Smith and Bruce R. McConkie and their more fundamentalist-leaning material. However, only some showed up for the discussion. Rodney Turner, who had not been attending, came in and threw down some creationist literature called "33 Questions That No Evolutionist Can Answer" and complained, "Why don't you guys look at this? You guys can't handle any of that," and said that it was for next week's discussion. Despite his bluster, he did not show up the following week. Jeffery was surprised by this level of contention and found it striking that so few were willing to examine their belief by directly confronting the evidence for evolution.[5]

Jeffery lamented that the seminar did not work out and that a reasoned discussion and airing of views had reached an impasse. Jeffery brought into the seminar much of the evidence he used for his "Seer" paper, which was met with skepticism and dismay from certain members of the religion department. The material he would soon publish was largely unknown and significantly complexified the portrayal of the church's position on evolution presented by most sources available on the topic at the time, especially Fielding Smith's *Man, His Origin and Destiny*. There is a temptation to see Jeffery as vindicated, casting this group of religious professors as benighted anti-intellectuals, but this would misread the intellect and intent of these men. Their stated position that evolution is of the devil is a response built on the works of men like Fielding Smith and other high-ranking leaders in the church, as sketched above. These religious educators were convinced they were doing the Lord's work by undermining the science of evolution. Like the New Testament depiction of the Apostle Paul, their zealotry was part of a love of the gospel and a reaction to perceived threats against it.

A year later, in 1973, almost all the religion professors ceased to attend the seminar, which had convinced few to modify their positions, nor had it ameliorated the intense negative feelings that each side of the debate harbored. However, there were a few who had not yet decided to take a stance and who held open minds that allowed them to reconsider their anti-evolutionary leanings. One was Noel B. Reynolds, at that time a professor in the philosophy department (later joining the political science department) and one of the founders and originators of the Foundation for Ancient Research and Mormon Studies (FARMS) which looked for evidence linking the Book of Mormon with modern archaeological studies. He recalled that in the seminar, "the religion guys were dead set against teaching evolution" and said positively about the seminar, "but it was, I thought, fantastic that all these hardliners on both sides were talking to each other. Jeffery was clearly, one, a faithful Latter-day Saint, and two, one of the best-informed people in the room, and the religion folks, by and large, really did not understand evolution, did not understand modern biology really well."[6]

He then notes that Jeffery was held in some suspicion by members of the seminar, who pointed out that because so many BYU students had been through seminary, they were apt to be against evolution. Reynolds then points out that many of the faculty attending the seminar wondered if Jeffery, and the other members in seminar who were supportive of evolution, were really good members of the church. In particular, they identified Duane "as somebody you have to watch out for."

However, Reynolds came away with a very different impression of Jeffery: "And it became clear to me in these discussions in the 70s that the biologists were being honest, and not rebellious. They were just honest inquirers, and the evidence was, by that point, just piling miles deep." When asked if the seminar helped, Reynolds said, "It did in a way, yeah. I don't think I depended on it in any deep way, but I was certainly impressed by it."[8]

That most of the religion faculty abandoned the seminar was not surprising, given that almost all of them embraced anti-science and especially anti-evolutionary feelings inherited by CES inculcation in anti-evolutionary influences. Duane Jeffery said that one of his students, who went on to join the religion faculty, commented about this period:

> "We are making progress over here." I said, "Like what do you mean?" and he said, well, when he joined the faculty, they would never have a faculty meeting but that some of these older brethren would stand up and often, with tears running down their cheeks, plead with the others and tell the others how important it was that they hold the line of the gospel against the atheist scientists on this campus, was the quote. And he said, "We may still have some that feel that way, but they would never stand up and say that."[9]

At the time, the contention between these groups was fierce. A sense that this was a continuation of the war in heaven between Satan and God was deeply entrenched in parts of the religion department discourse.

"Seers, Savants, and Evolution": An Overview

Jeffrey's work, "Seers, Savants, and Evolution: The Uncomfortable Interface," published in 1973, marked the beginning of his public defense of embracing organic evolution within the church. Jeffery begins his paper, "Ever since this great synthesis, Darwin's name has been a source of discomfort to the religious world." He opens this piece by establishing that Darwinian evolution is established science. He says in the 1981 Bergera interview,

> And then I try to characterize the way that was received and the fact that it was, it could not be, it was too well documented to be ignored.

> I then try to make that statement: The question of species is no longer open. *The problem with our science and religion literature in the Church is that most of it is not worth the time to read it. Anyone who's got any background in science knows that most of it is junk.* So, it was my aim in that first paragraph to let them know that this was going to be a different breed. And I think we were successful in doing that. But that first paragraph also turned off some of the anti-science people. I wasn't quite so concerned about them because I didn't think I was going to reach a lot of them anyway, who already had their mind made up. *The ones I would reach would be students.*"[10]

"Reaching the students" would become a dominant goal, the focus for the rest of his career. We will now look at the structure and content of the paper.

Structure

1. Introduction of His Aims and Purposes

Jeffery's first step is to lay out that most of the engagement within the church has assumed that evolution is a detriment to the church and can only harm it.

> We do claim, however, to try for the first time to document another, broader point of view fundamentally different from those which have been most ardently presented in the past twenty years and to examine in as complete a context as is currently sufficiently documented the statements of the prophets of the Church on the matter.
>
> Our account may be disturbing to some. It is not designed to be, but the nature and history of the subject make it virtually impossible to avoid affront to someone.[11]

Jeffery then points out the lack of actual engagement with the science of evolution by many members of the church. Although in the early church, there was an apparent belief that LDS were friendly with science and considered it one of their tenets to seek truth, regardless of its source. He quotes David O. McKay: "But science dominated by the spirit of religion is the key [to] progress and the hope of the future. For example, evolution's beautiful theory of the creation of the world offers many perplexing problems to the inquiring mind." But then Jeffery points out that "God's possible role in the voluntary processes has not been characteristic of Mormon literature" and explains that a possible reason "appears to be that LDS doctrines central to evolution were not well developed; they were still in a sufficient state of flux that no confrontation was possible or necessary. Simply put, the Church had no defined basic doctrines directly under attack."

2. A Comparison with Creedal Christianity and LDS Theology That May Be Relevant to Finding an Accommodation with Evolution

After acknowledging that members may find what follows disturbing, Jeffery attempts to alleviate concerns. He uses the six points of contention between Protestant Fundamentalist views and evolution, as outlined by evolutionary biologist and philosopher of science Ernst Mayr. Each of the following he critiques some of the status of science and then sees how most of these should not be a threat to LDS theology.

1. Belief in an ex nihilo creation, that is, creation out of nothing
2. The belief that the Earth was created in six twenty-four-hour days and is only about six thousand years old
3. Fixity or immutability of species; that all species were created initially in Eden by the Creator and did not change significantly
4. The contention that life depends on an activating vital force that is immaterial and divine—that is, spirit or soul
5. Special creation of man; God literally molded man's body from the dust of the ground and blew into it the breath of life, the spirit

He then takes these tenets and carefully dissects how these five points of contention are not at stake within Mormon doctrine.[12] By examining how the church's doctrinal tenets are sufficiently distinct from most forms of Christianity, he concludes that our materialist leanings more closely align with an evolutionary framework than with mainstream Christianity's ex nihilo creationist views. He notes that the last might be the most relevant to LDS. Still, he points out historical precedents where a literal interpretation of the "Creation of Man" above was repudiated, as seen when Brigham Young censured Orson Pratt for claiming that man was created from the dust of the Earth.[13]

3. Examines the Teachings on Evolution by LDS, Especially Its Leaders, Making a Historical Survey, Especially Noticing the Diversity of Opinions Among General Authorities About Evolution

For much of the rest of "Seers and Savants," Jeffery charts the position of several leaders of the church and their interlocutors. He carefully reads and critiques the 1909 statement by Joseph F. Smith, John R. Winder, and Anthon H. Lund, then the First Presidency of the church. Jeffery points out in some detail that although the 1909 statement can be interpreted as anti-evolutionary, the clarification provided as an editorial by joint editors President Joseph F. Smith and Edward H. Anderson in the April 1910 *Improvement Era* demonstrates the *evolution* of the church's position on evolution:

> Whether the mortal bodies of man evolved in natural processes to present perfection through the direction and power of God; whether the first parents of our generations, Adam and Eve, were transplanted from another sphere with immortal tabernacles, which became corrupted through sin and the partaking of natural foods, in the process of time; whether they were born here in mortality, as other mortals have been, are questions not fully answered in the revealed word of God.[14]

Jeffery uses this as a plausible argument for considering evolution as a possible interpretation for their bodies having been "born here in mortality." He also notes that Joseph F. Smith considered the matter closed after this statement.

4. Last, the Paper Looks at the Disagreement Among Several Church Leaders. Several Church Leaders and Historians, Including B. H. Roberts, Apostle and Theologian John Widtsoe, Apostle and Geologist James E. Talmage, and Apostle Joseph Fielding Smith

As detailed in the last chapter of this book, the rest of Jeffery's paper outlines the disagreements between church leaders and historians B. H. Roberts, Apostle and theologian John Widtsoe, Apostle and geologist James E. Talmage, and Apostle Joseph Fielding Smith. All of them had complex feelings on evolution, with Roberts and Widtsoe defenders of evolution, Talmage committed to evolution save in the case of humans, and Smith a staunch anti-modernist creationist.

Throughout his paper, Jeffery maintained a measured and respectful tone. Acknowledging that few members understood how there were disagreements within the church hierarchy and that the impression that the church was against evolution was primarily the result of the debate being decided by the death of all but Smith, and not any official revelation that evolution was false.

The paper was considered so controversial because it seemed to contradict several leading General Authorities who had spoken confidently that evolution was in direct contradiction with several key doctrines of the church. This idea was articulated by Apostle Boyd K. Packer in his talk, *The Law and the Light*, which purported to demonstrate that if the evolution were true, that the common interpretation of the creation wherein humans were created before other kinds of life could not be accurate, therefore the fall of man by eating the forbidden fruit could likewise not be true. If these two pillars of church doctrine were not true, there would be no need for the atonement of Jesus Christ; the central doctrine of the church would be obviated. He also argued that contrary to Jeffery's paper, which showed disagreement among the leading brethren about the possibility of evolution, the early leaders of the church were united against evolution. But Jeffrey's paper began to gain attention. From its publication onward, it was cited by many historians who corroborated the quality and accuracy of Jeffery's research, as well as by other scientists working to help members understand the importance and impact of evolution in the world of science. Students in the science classes widely read it. The paper was widely read and cited, becoming one of the most republished papers from *Dialogue: The Journal of Mormon Thought* ever published.[15]

The Reaction to Jeffery's Paper Was Swift and Brutal

During his sabbatical in Hawai'i to work on fruit fly genetics, Jeffery heard rumors that he might have been implicated in a devotional speech at BYU in which Ezra Taft Benson, then-president of the Quorum of Twelve Apostles, had attacked evolution. Four days after the speech was given on December 10, 1974, he wrote a friendly letter to Benson asking for a copy of his speech.

His letter was acknowledged, and Jeffery was sent a copy of the talk, "Jesus Christ—Gifts and Expectations."[16] The talk essentially looked at the role of General Authorities and declared their interpretation of doctrine to be definitive. Three points in the talk touched a nerve with Jeffery. The first was Benson's point that Christ gave us the Prophets, and we were bound to listen to them: "[Christ] has provided us the gift of a prophet. Of all mortal men, we should keep our eyes most firmly fixed on the captain, the prophet, seer, and revelator, and President of The Church of Jesus Christ of Latter-day Saints. . . . The inspired words of the president are not to be trifled with. . . . *Therefore, the most crucial reading and pondering which you should do is the latest inspired words from the Lord's mouthpiece.* That is why it is essential that you have access to and carefully read his words in Church periodicals."

The second concern Benson addressed was the proper role of the church in arbitrating truth claims and how those members who question things are likely to be trying to destroy the church. "Because God has given men their agency, there will always be those who will misuse it. The gospel net draws in the good and the bad, the best and the worst—the worst because the devil, before the final cleansing, puts some of his followers within the kingdom to try and destroy it. We have some of them within the kingdom today, and in due course their number shall be known." Jeffery considered this quote a direct and unmistakable reference to himself.

Finally, Benson discussed how the Book of Mormon, correctly interpreted, will help subvert the philosophies of men: "If we really did our homework and approached the Book of Mormon doctrinally, we could expose the errors and find the truths to combat many of the current false theories and philosophies of men, including socialism, humanism, organic evolution, and others."

Jeffery recognized this as a personal attack on his work, so he wrote a thoughtful letter to President Benson to question what the Apostle had put forth in his talk. Benson's reply, a single short paragraph, admonished him to read Fielding Smith's *Man, His Origin and Destiny*. Jeffery replied with some frustration:

> I have read Brother Smith's book, quite thoroughly, and at least several times over the last fifteen years. I have pursued in considerable detail the reference materials he cites, both scriptural and otherwise, and have pursued the teachings and counsel of others of our prophets as well.

> Indeed, it is the result of all that which prompted my letter to you, with some distilled questions—which still stand.[17]

President Benson's secretary then took over, admonishing Jeffery against evolution by pointing out that his stance on evolution contradicted fundamental gospel doctrines. He then included a letter from a biologist, Stephen L. Clark, in Ogden, who has suspicions about the scientific validity of organic evolution. Jeffery responded to Benson's secretary with some thoughts on his personal beliefs about why evolution matters.

The conflict between Jeffery and Benson escalated at a fourteen-stake fireside[18] where President Benson gave a talk celebrating the two-hundredth anniversary of the founding of the United States. Benson was an active presence in the John Birch Society, a reactionary right-wing organization that actively taught that communists were infiltrating all levels of American society, and only extreme vigilance would keep the United States from falling prey to these machinations. Although Jeffery did not have access to the text of the actual talk, he noted that the nearly identical talk from March of that year, "God's Hand in Our Nation's History," gives the flavor of the given speech in which Benson explores the founding of the United States through the lens of the Book of Mormon prophecies about its place in world history. Benson begins, "Tonight, I will speak to you about our beloved republic and the inspired agents whom God raised up to establish the foundation upon which our liberty rests. I will speak to you also about some mischief that has been afoot for a number of years, a mischief that intends to undermine our republic, its founders, and the Church."[19]

Benson decries secular histories that do not honor the founding fathers or that dismiss God's hand in the country's "origin and destiny," then calls out people criticizing church leaders like President Brigham Young. Then abruptly, he shifts gears and begins a direct attack on Jeffery's *Dialogue* paper:

> More recently, one of our Church educators published what he purports to be a history of the Church's stand on the question of organic evolution. His thesis challenges the integrity of a prophet of God. He suggests that Joseph Fielding Smith published his work, *Man: His Origin and Destiny*, against the counsel of the First Presidency and his own Brethren. This writer's interpretation is not only inaccurate, but it also runs counter to the testimony of Elder Mark E. Petersen, who wrote this foreword to Elder Smith's book, a book I would encourage all to read. Elder Petersen

> said: Some of us [members of the Council of the Twelve] urged [Elder Joseph Fielding Smith] to write a book on the creation of the world and the origin of man. . . . The present volume is the result. It is a most remarkable presentation of material from both sources [science and religion] under discussion. It will fill a great need in the Church and will be particularly invaluable to students who have become confused by the misapplication of information derived from scientific experimentation.
>
> When one understands that the author to whom I alluded is an exponent of the theory of organic evolution, his motive in disparaging President Joseph Fielding Smith becomes apparent. To hold to a private opinion on such matters is one thing, but when one undertakes to publish his views to discredit the work of a prophet, it is a very serious matter. It is also apparent to all who have the Spirit of God in them that Joseph Fielding Smith's writings will stand the test of time.

Benson then goes on to address the possibility that his take on this matter might be considered inappropriate for the venue at which he is speaking:

> I suppose there will be some who will suggest that a fireside is neither the time nor the place for these kinds of remarks, that they would be better confined to the closed door of a faculty forum. My purpose this evening is to help you to discern a trend that has been destructive to the faith of many of our people in our nation's founders and our country's divine origin and destiny. My purpose further is to forewarn you about a humanistic emphasis which would tarnish our own Church history and its leaders.

The night after the talk, ten students came to Jeffery's house. They said, "Boy, did you get it tonight!"

"What are you talking about?" Jeffery answered, this being the first he had heard about the talk. He goes on, "I spent much of the next day on the phone with various people calling in, and Harold Bissell from geology was fun and games (he was known as a heretic anyway), 'Well, you finally got their attention! The rest of us have been trying for thirty years!'"[20]

Jeffery was deeply affected by his being so explicitly called out by one of the Apostles of the church. His relationship with General Authorities had been one of respect, even while noticing and acknowledging their foibles. He took the threat implied by Benson's talk very seriously. He wrote a carefully worded letter to the university's president, Dallin H. Oaks. His fear and concern are apparent, and he refers to Benson's talk as "charges" and spends a considerable amount of ink trying to defend his work on evolution and the church and on his resolve to stand by the things he has written.[21]

In September 1977, as a second attempt to bring the subject of evolution into a conversation among the various faculty, the College of Physical and Mathematical Sciences and the Religious Instruction faculty organized a set of one-hour seminars for faculty, interested graduate students, and advanced undergraduates to discuss evolution and related topics. Things, again, did not go well. Bergera relates that one of the participants had written to President Benson, raising concerns about Jeffery's inclusion as one of the speakers. After the first two sessions, William O. Nelson, secretary to the Quorum of the Twelve, complained to Elder Benson that the program was pro-evolutionary and was angry that Jeffery had been allowed to speak giving a sense of how Jeffery was perceived by some of those threatened by his work on suggesting that the church and evolution could be reconciled productively.[22]

Nelson's letter also details his understanding of how BYU evolutionary biologists approach faith and science, which he felt undermined student faith and dependency on the Savior. "That to me is the tragedy. It seems to me that this science and religion seminar is a good case in point that evidence is your concern with the slanting of the courses at Brigham Young University." He then adds, "Where are the voices of testimony? Where is the strong declaration to the students of the position of the Church on these matters? I am confident it is not going to come from any of the speakers that are represented on this panel."[23]

Nelson's last paragraph in the letter shows that anyone holding views favorable to evolution did not have a testimony, or at least a proper one. It is striking in these exchanges that Benson viewed the Seventh-day Adventist creationist-informed writings of Fielding Smith as canonical. This was the underlying assumption of many within CES, and with the president of the Quorum of the Twelve Apostles actively promoting that idea and directly attacking proponents of evolution, it was, despite Jeffery's important historical research, an uphill battle trying to persuade members of the church that the church had no stance on evolution.

For example, two years later, in August 1979, Julian R. Durham, a World War II veteran and U.S. government worker, wrote a letter of complaint to Ezra Taft Benson about instances he noted where evolution was being sneaked into the Saints' discourses.[24] The first he mentions is when Spencer Kinard, in his *Music and the Spoken Word* broadcast, stated, "Great mountains being millions of years of age." According to Durham, this is "a false statement and diametrically opposed to the scriptures, in the Standard Works, as you well know." He then cites Fielding Smith's *Answers to Gospel Questions* and *Man, His Origin and Destiny* as evidence that evolution and talk about ancient Earth are anathema. He then tells the story of his son-in-law having to stand up to the truth as a BYU religion professor ridiculed his oral report on the age of the Earth based on Smith's compilation based on the scriptures. "Aren't you ashamed of this type of thing taking place at the BYU?"

He rants a bit about the state of education in America after stating that he wishes that "President Kimball would plead to the Lord about the BYU situation and get an answer from Him as he and the brethren did on the black problem" (a reference to the Priesthood and Temple Ban that restricted the priesthood and temple blessings to people not of African descent). He then adds, "You know, as President Kimball, our prophet knows, that the stench of apostasy is permeating many departments on campus."

At the end of the letter is a handwritten "PS" complaining that the director of the Earth Science Museum, James "Dinosaur Jim" Jensen, is a devotee of organic evolution and had publicly declared that a fossil bird he discovered was 140 million years old. "What nonsense!" Durham declares.

This letter is fascinating in exposing how the leading brethren of the church took seriously input from well-heeled members of the church who held worldly rank and position. The letter was passed to Gordon B. Hinckley, who passed it to the commissioner of education, Jeffrey R. Holland, with the note dated August 29, 1979, "I think this should be placed on the agenda for the next meeting of the Executive Committee and that President Oaks should be given an opportunity to see it in advance so that he may come prepared to comment on it."[25]

President Oaks's detailed letter responding to President Hinckley is instructive because of his defense of BYU and its faculty.[26] It is especially telling that he directly defends the teaching of evolution as a necessity to understanding biology and that the students need to understand this if BYU is to continue its function as a university. There is no attempt here to blame or correct the faculty and a strong defense of their continuing tolerance for teaching evolution as they have in the past.

This suggests one of the most challenging tasks of the Church of Jesus Christ of Latter-day Saints. As in other places, President Benson outlined in his talk, "Jesus Christ—Gifts and Expectations,"[27] that the church will not be led astray by its leaders, and in his fireside comment attacking Jeffery in unmistakable gestures to his work, "when one undertakes to publish his views to discredit the work of a prophet, it is a very serious matter."[28] This became part of the reason that arguing that the church has no stance on evolution continues to meet pockets of resistance from members of the church and in CES in particular. Although recent attempts have added nuance about the fallibility of the men who lead the church,[29] in general the church has done a poor job of making it safe to disagree with leaders on several scientific and social issues.

Jeffery felt leader worship continued to be a problem in the church where almost all instruction in the church, including lessons to accompany CES manuals and those used in Sunday school for weekly instruction, contain a significant amount of repeated discourse from General Conference talks from the leaders. Much of the twentieth and twenty-first century has been spent in trying to navigate the modern church through contemporary challenges such as race and the civil rights movement, the role of women, the church's response to queer members, and LGBTQ+ relations.

This is evident in the relationship between Fielding Smith and evolution; his knowledge about the issue was piecemeal at best and biased by those arguing for a young Earth in particular. Whereas much of his other work has been essential in creating a commitment to gospel scholarship, and he is remembered as one of our important gospel scholars.[30] Nevertheless, his work on creation has been problematic for the church relative to the most basic commitments of biological science. It is interesting that after he became the president of the church, as far as could be found after an exhaustive search, Fielding Smith did not talk about organic evolution or Darwin again. In the same year he was made the president, he did say in the 1970 October general conference, "What I have taught and written in the past I would teach and write again under the same circumstances."[31] Although not a repudiation of this thought on evolution, neither is it a specific reference to it, and the word "circumstances" might be taken as an acknowledgment that the "times" in which he lived informed his teachings and writing. Publicly, he never mentioned the subject of evolution directly as church prophet.

As all people's perspectives are colored by the times and places of their upbringing, choosing to prioritize the protection of mistaken ideas as a way of guarding people's reputations becomes problematic. Most humans hold opinions that will be corrected, modified, nuanced, and discarded by future scientific perspectives. Discovering problems with opinions held by people who have become enshrined as irreproachable can cause more problems than necessary for thoughtful members. In his 1981 interview with Gary Bergera, Jeffery relates these concerns, which were very apparent to him then:

> I think our current picture of the infallibility of the brethren cannot be tolerated. Not that I want to break it down, but because it throws us into a way of thinking that then traps us. And I think that's dangerous. Now, I can appreciate why the brethren particularly all through this century have labored extensively to keep up the image of unity among the brethren. But I am afraid that while that image is valuable, it is costing us dearly. And I personally see no reason to continue to promote that at quite the level we do.[32]

Jeffery seems to anticipate what has become a very pressing problem in the contemporary church, often thought to reflect the amount of historical information available from the proliferation of the internet and social media, but that he recognized clearly in 1981:

> It's been my experience in the Church that the most difficult kids to keep in the Church are those who realize they've been lied to with our Church history and it usually is not any great problem to help them understand what went on, say, in the nineteenth century. What is tough is to get them to accept the lying that has gone on directly to their face, in some instances in the twentieth century. And then ask them to sustain as Church leaders and so on some people whom they know are deliberately prevaricating.[33]

Following this peek into what was going on at the highest levels of the church concerning the teaching and study of evolution at BYU, it is worth a look at what was happening on the subject among the disparate faculty there. That there was confusion about the church's stance on evolution is unsurprising, given that some of the most prolific General Authorities denounced it. This was especially true in CES, including the information provided by the church for its Seminary and Institute program for the education of young people. In Utah, students were given release time from their high schools to attend such classes. In other parts of the world, students often gathered in classes held early in the morning before their secondary education classes started. Institute was mainly for college students and provided continued instruction in scripture study, a version of Mormon doctrine, and history. In the next chapter, we will explore some of the current manuals. Additionally, there have been negative associations about evolution from the pulpit, both in print and in oral presentations like firesides, stake meetings, and general conferences.

In 1979, Jeffery and a group of scholars from several scientific disciplines published a two-volume book exploring how religion and science fit together, called *Science and Religion: Toward a More Useful Dialogue.*[34] As William Evenson notes in a 1980 review of the project in *BYU Studies*, the project was started seven or eight years before the actual publication, and participation by BYU faculty was approved by the BYU administration, the Church Commissioner of Education, and the Executive Committee of the Board of Trustees.[35] It was initially to be published by a church-affiliated press like Deseret Book, Bookcraft, or BYU Studies. However, the work experienced several delays, and in the end, the project did not seem financially viable, so a non-LDS publisher eventually published it, giving it limited distribution among members of the church.

The first volume contained largely philosophical articles, along with works looking at the dating methods used to estimate the age of the Earth, and other papers focusing on the physical sciences. The second addressed some difficulties people often encountered in reconciling faith with evolution, as explored by many believing LDS anthropologists, archaeologists, physicists, biologists, and geologists in the special sciences. William Evenson reviewed the work, and despite finding unevenness in the level of technical expertise needed to understand some of the papers and despite some articles veering into poor apologetics that bordered on fundamentalist arguments, his overall assessment was that this was an excellent look at the challenges and opportunities of science and religion dialogue. Unfortunately, the work never entered the mainstream consciousness of LDS discourse, whether because of the difficulties Evenson lays out or because it was published by a non-LDS press.

The book is a valuable demonstration of interest in conversations about how science and religion fit together, how the two differing perspectives can be harmonized, and how Mormonism can embrace both ways of knowing.[36]

In February 1982, on Utah's PBS show *Civic Dialogue*, Duane Gish, the associate director of the California-based for Creation Research, debated Duane Jeffery.[37] Gish was a well-known creationist biochemist who traveled the country publicly debating evolution versus young-earth creationism. He was well prepared for typical scientific responses and had a reputation for leaving the evolutionary scientists he debated feeling unprepared and having lost the debate. The *Washington Post* ran a piece about Gish and looked at a debate he had with University of California biochemist Russell Doolittle.[38] Doolittle was trounced. Gish specialized in embarrassing evolutionary biologists who arrived thinking that with the facts on their side, what could go wrong? This became such a problem that Catholic evolutionary biologist Kenneth Miller, author of *Finding Darwin's God*,[39] wrote a piece for the National Center for Science Education's *Creation/Evolution Journal* to help evolutionists debate against such interlocutors as Gish.[40]

A group of students watched the Jeffery-Gish debate a few months after the debate aired. One student remembers that after it was over, there were fist bumps, wild clapping, and cheering and that Jeffery had "smoked him." The student also remembers saying that a general feeling of elation accompanied the event.[41] Videos of such debates can be found on the Creation Research Institute's website but, unsurprisingly, not this one. Part of the reason that Jeffery was so effective against Gish was that the creationist evangelical was not used to debating fellow religionists, especially one that was a bona fide evolutionary biologist and geneticist. Jeffery, as a fellow believer, was able to skirt many of his standard tactics and strategies. Most of Gish's audiences were Christian believers. To his surprise, Jeffery knew the biblical story of Genesis as well as he did and could offer counterarguments from the Bible itself.

Not all were as pleased as this group of undergraduate students. Angry letters poured into BYU President Jeffrey Holland's office, complaining about the BYU professor who had argued the wrong side of the debate. Jeffery said that Holland called him into his office in reaction to the letters from parents of BYU students who were disturbed by his performance with Gish, the BYU president was upset and said that he "didn't want to have to announce machine-gun Jeffery at sunrise by the flagpole."[42]

After the school's paper, the *Daily Universe*, ran a story called "Evolution, Creation," which quoted Jeffery saying that students did not "have to adopt the very strained anti-science stance that so many have been indoctrinated with," President Holland met with the chairs of the biological and agricultural sciences department chairs, expressing concerns that the mission of BYU had been compromised by biology professors speaking publicly about evolution. Jeffery was asked for proposals on how controversial subjects could be handled in "ways that would be supportive and not additionally inflammatory." Jeffery responded, "We really can't avoid the topic [of evolution] as it is central to our discipline. We sincerely hope and pray that the day may come when we can speak freely without fear of being misunderstood."

A year after these events, Jeffery would publish a book review in evolutionary biology's most prestigious journal, *Evolution*, about how to engage with creationists. Beginning with the Gish debate, Jeffery would continue to actively combat the influence of creationists for the rest of his life. It was about this time that Jeffery's focus began to shift to an academic interest in science education as a discipline.

Jeffery's paper opened opportunities for evolution to be discussed at multiple levels within the church hierarchy, at BYU, and throughout the church. In the next chapter, we will look at further attempts to educate the Saints about science in general and evolutionary science in particular.

CHAPTER FOUR

Educating the Saints on Science

Jeffery and his associates in biology and its related disciplines like geology saw the rise of Protestant Fundamentalism as the next challenge in the science education of the Saints. This brand of fundamentalism, with its roots in the late nineteenth century, increasingly influenced the discussions in the United States, ranging from what should be taught in schools to the way that government entities operate in their treatment of religion.

Jeffery was uniquely positioned to take on this challenge, as he was known to be academically rigorous, informed, honest, straightforward, and unafraid to stand up for the truth regardless of the power structures that might differ from his views. His integrity was beyond reproach. He was knowledgeable about and committed to the church and its teachings. He also had a strong commitment to building and strengthening his community. This would inform his interaction with the Saints throughout his life.

These attributes come into play in his steps to respond to the concerns over fundamentalism, and he was determined to help the members of the church understand that the kind of creationism being taught by other Christian churches was not the same as the story of the Creation taught by the Church of Jesus Christ of Latter-day Saints. This established four important priorities that defined the remainder of Jeffery's time at BYU: (1) working with students to help them grapple with the anti-science sentiments picked up in Church Education System (CES) classes influenced by the anti-modernist, anti-science, and anti-evolution views held by some important leaders of the church; (2) communicating to the Saints the nature of the creationist fundamentalist creep into the doctrines of the church by confronting that influence directly; (3) fighting to keep the Utah State

Legislature from introducing legislation demanding secondary school teachers spread ideas from Protestant Fundamentalist creationism into the classroom, especially through the clever marketing ploy called "intelligent design creationism"; and (4) helping people understand science through his long-running column, "Science and Society," which ran as a weekly from 1990 to 2015 in the *Provo Daily Herald*.

Working with Students to Help Them Reconcile Their Faith with Science

Jeffery's integrity and faith show up clearly as he helped students reconcile the church's doctrines with science. From the earliest days of the school's history, letters from students and their parents to BYU administrators and general leaders of the church have challenged the school's leadership. As noted in chapter 2, much of the 1911 crisis at BYU was exacerbated by letters from concerned parents. They responded, in part, to the lack of clarity about the church's stance on evolution, and the diversity of opinions of vocal general authorities led many of the Saints to come away confused about how to understand this controversial topic. Moreover, for many years, the official church manuals clung exclusively to the thinking of some of the leaders with negative perceptions about evolution, who often championed a negative view of science in general. Because the manuals designed for Institute (the church's education for college students and above) and Seminaries (church education for children of high school age) frequently contained quotes and perspectives that cast suspicion on the value of teaching organic evolution and other sciences, many parental letters expressed concern for their children's exposure to organic evolution.

These anxieties about the place of evolution in the curriculum at BYU dogged Jeffery throughout his career, including having to address concerns from parents, colleagues (especially members of the religion departments), and other institutional voices. Many of his most challenging confrontations orbited his insistence that the church had no stand on evolutionary science—a stance constantly undermined by content within the church's internally published materials.

Consider the rhetoric of the CES manuals used for religion classes at BYU during Jeffery's tenure (with similar manuals also being used in Seminary and Institute classes) that reference the negative views of science found in the anti-modernist work of Joseph Fielding Smith. For example, the CES manual on the Old Testament raises suspicions about how science has dated the age of the Earth. Here is the text examining Genesis from the Hebrew scriptures:

> Even when it is realized that chapter 1 of Genesis does not describe the beginning of all things, or even the starting point of mankind, but only the beginning of this earth, it cannot be said definitively when that beginning was. In other words, the scriptures do not provide sufficient information to accurately determine the age of the earth. Generally speaking, those who accept the scriptural account subscribe to one of three basic theories about the age of the world. All three theories depend on how the word "day," as used in the creation account, is interpreted.[1]

It is telling that after admitting that "the scriptures do not provide sufficient information to accurately determine the age of the earth," the passage does not point out that good answers to the question have been found through scientific investigation and used effectively for nearly a hundred years. The manual intimates that the question is only accessible through scriptural readings or direct prophetic revelation. The author of the manual approaches the question of the age of the Earth by handwringing about various interpretations over the use of the word "day." Students are not directed to any scientific evidence found in the disciplines of geology or physics, nor does the passage offer any scientific sources that might be used to figure out the question. This leads students to the impression that revelation is the only source relevant to the question and obliquely casts suspicions on science as a way of explicating the world's natural processes.

At the end of the chapter in the CES manual is a section called "Points to Ponder," where the anti-science stance is taken further. The manual gives six points from Genesis on the Creation that explore implications for LDS belief, all of which hew closely to the 1909 statement on evolution and focus on some elements that seem to target Darwinian evolution as an unacceptable theory. Number six says, "The world was not created by chance force or random accident."[2] Like most creationist literature, this significantly misunderstands how randomness functions in evolution. Earlier in number five, the manual notes, "Adam fell from a state of innocence and immortality, and his fall affected all life upon the earth as well as the earth itself." This plays on Fielding Smith's syncretic reconstruction of LDS theology based on the Protestant Fundamentalist young-earth creationist idea that there was no death on Earth before the fall of Adam.[3] These subtle hints of anti-science become much more obvious, and Darwinism is brought into the discussion of Genesis explicitly. This paragraph introduces the section:

> In the world another theory of how things began is popularly held and widely taught. This theory, that of organic evolution, was generally developed from the writings of Charles Darwin. It puts forth different ideas concerning how life began and where man came from.[4]

In a string of quotes from various leaders of the church, including this statement from McConkie's compilation of Fielding Smith's thought in *Doctrines of Salvation*, Darwin's evolution is explicitly attacked:

> Of course, I think those people who hold to the view that man has come up through all these ages from the scum of the sea through billions of years do not believe in Adam. Honestly, I do not know how they can, and I am going to show you that they do not. There are some who attempt to do it, but they are inconsistent—absolutely inconsistent, because *that doctrine is so incompatible, so utterly out of harmony, with the revelations of the Lord that a man just cannot believe in both.*
>
> . . . I say most emphatically, *you cannot believe in this theory of the origin of man, and at the same time accept the plan of salvation as set forth by the Lord our God. You must choose the one and reject the other, for they are in direct conflict and there is a gulf separating them which is so great that it cannot be bridged*, no matter how much one may try to do so.[5]

Here, we find a clear statement of opinion by a leading general authority of the church that belief in evolution and in the church are incompatible. The manual then refers to creationist Harold Coffin at Seventh-day Adventist Andrews University and his 1975 book *Creation: The Evidence from Science*,[6] which misrepresents the state of science so egregiously it becomes a form of pseudoscience.

The manual then dismisses evolution and claims that Creation is the best way to interpret the fossil record. While typical of creation literature, this certainly does not reflect mainstream science or other religiously faithful responses, which argue that Darwinian evolution and Creation are compatible. If promoting the best scientific thought and practices was an essential part of restoration thinking, quoting such a source that contradicts a united scientific consensus about the centrality of evolution in the biological sciences and scores of related disciplines undermined the purposes of Church Education. Jeffery would often note in his classes the scripture from Doctrine and Covenants 88:118: "And as all have not faith, seek ye diligently and teach one another words of wisdom; yea, seek ye out of the best books words of wisdom; seek learning, even by study and also by faith." This perspective carried with it a sense that one of the background assumptions of the LDS faith was that learning and knowledge were desirable. "The glory of God is intelligence," from Doctrine and Covenants 93:36, was used as a stock phrase in much of BYU's discourse and as an explication of its aims.

I have used these extensive quotes to show how this information in CES manuals created conflict for students of science and other disciplines as they tried to reconcile their faith with science. The quotes above are all available in current (2023) manuals and, as of this writing, are still present on the church's website. Nowhere in the manual could I find any science-positive statements, save that "true" science and religion will not conflict (with "true" marking a suspicion that contemporary science has moved away from where it should have gone). However, the suspicions raised here provide a sense that the conflict between science and religion is persistent and real. Church Education manuals since the 1950s have been remarkably anti-evolution, encouraging the view that the scriptures are the only source of valid information about the creation of the universe. The manuals also include only fringe pseudoscience and no examples of actual scientific research and argument. This often suggests to students that science should be approached with suspicion and promotes the idea that consensus among scientists is untrustworthy and that scientific conclusions should be held at arm's length. Such misreading does damage to science students' perceptions of their faith tradition, suggesting to their minds that if the LDS Church is committed to such shaky foundations as those seen in the creationist literature for their inferences about the nature of life on Earth, what else might they be wrong about? Counter to their intentions, CES may create suspicions about the faith rather than the science. Jeffery worked hard to dissuade students from this kind of thinking and tried to help science students to maintain their faith and to not throw out the proverbial baby with the bathwater. However, at times the pushback followed the opposite pattern. Some students ended up keeping the anti-science bathwater to support their perception of the church's stance on evolution and abandoning science. This plays out in significant ways as the following experience demonstrates.

Pushback from Students

In June 1988, one of Jeffery's premed students, disturbed by his teaching of evolution, wrote a letter of complaint to several leaders at BYU, including Bruce N. Smith, Jeffery's dean, and BYU President Jeffrey R. Holland. Jeffery notes concern that the letter was being passed around among other students in his class.[7]

The student begins his letter by explaining his extraordinary experiences at BYU and its classes and detailing his student successes, including his work as a research and teaching assistant and his high 3.82 GPA (he got a B in Jeffery's class, which might have lowered that a bit). He then says, "Dr. Jeffery [*sic*] in my estimation is atheistic in his teaching, for I have not heard otherwise from his lips. His entire purpose, from what I could gather, was to drive home that mankind has evolved from other lower primates. Gospel principles were seldom if ever brought up or discussed. The only time Creation was discussed was in jest when Dr. Jeffery referred to the 'idiotic ideas' of Henry Morris [a prolific young-earth creationist]." The student goes on to claim, contrary to fact, that most states demand equal time be given to creationism and evolution in college coursework. He then adds, "There are other concerns that I have regarding Dr. Jeffery's mental stability. I have talked to another professor about him, and he seemed to concur with my concerns." He then details an encounter where he was treated rudely when he came to Jeffery to complain about a test question in order to raise his grade. The student then suggests that evolution should be taught by other people, presumably in a way that better aligns with the student's creationist ideas.

To defend himself, Jeffery sent a lengthy response to Dean Smith and Jeffrey R. Holland, copying Provost Jae Ballif. He details false and misleading claims from the student and gives more context for his class and what it covers about the doctrine of creation, which was the focus of many of the student's concerns.

Like many professors at BYU, fear about accusations of "unrighteousness" from others are a source of fear of dismissal or even punishment. This is true today, especially after the purges in the early 1990s enacted in the events around the persecution of those whom Elder Packer infamously deemed enemies of the church, "'so called 'intellectuals,' feminism, and the 'gay-lesbian movement,'"[8] that resulted in the excommunication of the September Six (see Abbott[9] for a firsthand account of these events at BYU). The letter also represents a look at the concerns that Jeffery was haunted by in his continued attempt to navigate a commitment to science and evolution in the face of many members' growing unhealthy skepticism about science.[10]

Jeffery begins his letter by addressing the student's charge of rudeness, which related to the student having come to argue for more points to raise his grade. Jeffery remarks in his response, "If memory serves me correctly . . . he came to ask for additional points on the exam. His interpretation of the words used in the question were, in my view, novel indeed, and would, if accepted, have necessitated my recalling exams from all other students for re-grading—something I have tried in the past with rather poor success."[11]

To the claim that he did not bring up the gospel or creation in his class, Jeffery, who kept detailed notes after each class, provides his notes about specific dates and what was talked about specifically to counter the students' claims and which give an interesting illustration of what he was teaching: sensitivity of the issue of evolution to many members of the church; creationism defined; and the Supreme Court's decision on creation in the classroom; the church's pluralism on evolution; and an invitation for the students to read his "Seers, Savants and Evolution: The Uncomfortable Interface;" the nature of science; standard Christian explanations on age of Earth; Darwin's life; pluralism of LDS views, including Joseph Fielding Smith and David O. McKay; Henry Morse's arguments for a young earth; and a look at differences between creation versus creationism. He argued that he opened the class to students so they could explore their own questions.[12]

Jeffery then returns to the student's claim that he and an unnamed professor felt that Jeffery had issues of "mental stability." Jeffery states, "I'm at a loss here; this seems such a strange charge."[13] He then wonders if it is from a religious professor who has been trying to get rid of him: "My only possible insight here relates to a prominent religion professor, son of a former General Authority of Scotch ancestry, who has been reported to me on two occasions by students that he considers me to be 'possessed of Satan.' Is this new charge [of mental instability] just a more sophisticated version?"[14] Jeffery then laments the state of things he often faces while teaching at BYU, especially considering recurring challenges like this student represents.

> But in the past 3+ decades we have seen in the Church the growth of an anti-science sentiment, as is clearly documentable. Evolution has been equated by some with Korihor. Joel 2:28–31 has become almost a forgotten scripture and has not been publicly used for years in the sense above—and science, rather than being painted as a result of God's pouring out His spirit upon all flesh, is seen as an expression of the anti-Christ. That is most unfortunate, since, in my view, not only does it rob us of one of the critical aspects of Moroni's divine message, but it severely compromises us in our ability to legitimately address the world's problems. I will not detail the many medical, agricultural, natural resource management, technological and demographical problems to which evolutionary science has immediate and critical application. They are obvious to any person acquainted with the subject.[15]

Jeffery sees the increasing anti-evolutionary stance by many in the church and its leadership as evidence of a growing and concerning trend of anti-science in the church.

Countering Protestant Fundamentalist Creep in CES Directly

Because he was knowledgeable about and committed to the church and its teachings, Jeffery became involved in several public efforts to increase the Saints' knowledge directly, the *Encyclopedia of Mormonism* and Evenson and Jeffery's book on the official teaching from General Authorities about evolution.

The *Encyclopedia of Mormonism*

The *Encyclopedia of Mormonism* was to become a compendium of knowledge about the history and doctrines of the church.[16] Its acknowledgments section gives a brief history of its genesis:

> When Macmillan Publishing Company asked authorities at Brigham Young University whether they would be interested in developing an encyclopedia about The Church of Jesus Christ of Latter-day Saints, President Jeffrey R. Holland took the query to his Board of Trustees.

> They instructed him to proceed. Working closely with Church authorities and Macmillan, President Holland chose an editor-in-chief and a board of editors.[17]

The person assigned to write an entry on evolution was William E. Evenson, a physicist who served as academic vice president from July 1985 to August 1989. In general, the entries were by appropriate scholars with little input from official church sources. In the case of the entry for evolution, however, "little input" is not an inaccurate description of what happened.

Evenson came to BYU about the same time as Jeffery and moved into the same ward. He had been interested in biology for a long time, especially evolution, as can be discerned in his review of the 1979 response of scientists to issues about evolution and the age of the Earth. In 1977, he, like Jeffery, went to the University of Hawai'i for a sabbatical in the botany department to work on physics applications in ecology. He worked as the dean of general education for a time and then became the academic vice president in the summer of 1985.[18] While vice president, he worked with Elders Oaks and Holland on the *Encyclopedia of Mormonism*, so when he left administration in 1989, he was earmarked as writer for the evolution entry. He was asked to write a four-thousand-word entry,[19] but "the two significant editors dealing with that article both had robust ties to religious education, and not much science, and they didn't like it all."

Robert J. Matthews was the dean of religious education at BYU for eight and a half years and one of the associate senior editors on the encyclopedia. He had strong concerns about Evenson's initial draft—especially that it did not reference the works of several "notable General Authorities" including Fielding Smith, Ezra Taft Benson, Bruce R. McConkie, and Boyd K. Packer. He also wrote that Boyd K. Packer's anti-evolution paper, "The Law and the Light," should be included in the article. The depth of his concern shows as he writes in a July 16, 1990, memo to Noel B. Reynolds, director of the Foundation for Ancient Research and Mormon Studies and a member of the political science department and who served on the editorial board, that he wants to make sure the leaders of the church review the entry, stating, "There is little doubt but what the Advisors and others of the Brethren will want to read this article before it is printed in the *Encyclopedia*."[20]

Jeffery weighed in, giving Evenson some advice in a letter dated November 12, 1990, providing some information on what various General Authorities thought about the age of the Earth: "BRM [Bruce R. McConkie] eventually gave up on the age of the earth, saying that the days of creation were long eons, etc., though he still insisted that species cannot change even though we have been making them artificially since 1927. JFS [Joseph Fielding Smith] would never yield on the age of the earth. ETB [Ezra Taft Benson] has never discussed it. BKP [Boyd K. Packer], in his 1988 speech, was quite conciliatory for evolution of everything but humans; a position that Joseph Fielding Smith, Bruce R. McConkie, Mark E. Petersen could have never countenanced—Ezra Taft Benson never addressed the issue. One cannot claim, as RJM [Matthews] does, that 'they were all of one mind' when clearly, they were not."[21]

The efforts to establish a middle ground continued, but Matthews, whose stance against evolution was well known, became even more concerned that the entry on evolution was going astray because it did not reference anti-evolution statements by Apostles, as noted above. In 1988, he gave a talk to the religion faculty that stated that evolution was a moral issue because it was a subtle attack on the mission of Jesus Christ. A year later, condemning theistic evolution in a speech as the dean of religious education, he asked, "Do we rationalize evolution, and the creation, by the theories of mortal fallen men, stretching the revelations here and compressing them there?"[22] T. Benjamin Spackman's dissertation (2024) on Fielding Smith's creationist ideas identifies Matthews, who served for decades on the church's Correlation Committee and Scripture Committee, as one of the main influences in bringing Fielding Smith's views into the Church Education curriculum.[23]

Finally, because progress was stalled on the entry, Noel Reynolds sought help from Elder Dallin H. Oaks, then a member of the Twelve Apostles, as Matthews and the editor in chief, religion professor Daniel H. Ludlow (1924–2009), thought the evolution article was too pro-evolution. Oaks apparently had made comments on the article, but the editors were still concerned: "After receiving Brother Ludlow's report, we think it is important to make clear that neither of us carries a brief for evolutionary explanations of creation."

Matthews continued to find the Evenson article unacceptable, adamant that it did not reflect the brethren's view on evolution. Matthews decided to rewrite the entire article, constructing his interpretation using long quotes from General Authorities and insisted that his version reflects the proper stance to evolution and was the only correct one, evincing absolute certainty that Evenson was wrong.

Evenson's efforts to mollify Matthews, whose interpretation closely followed Fielding Smith's writings, reflected the notion that any positive mention of evolution subverts the church's doctrines about the Fall, the Atonement, and the Resurrection. In a memo to Daniel H. Ludlow, Matthews comments that Evenson's current version was "quite Telestial in quality," referencing the lowest kingdom in the LDS postmortal future of all living humans. He complained that "organic evolution as a process for the *origin* of man is not acceptable."[24]

In my interview with Evenson, he describes how the many versions he had written, along with Matthews's were given to President Hinckley by Elders Oaks and Maxwell:

> President Hinckley basically had his own version, but it included some of the things that I'd written. It was very short, as you have seen, and simply leaves things pretty open. So, then the editor called me and said, "Well, we've got this, and it's not what you gave us at all. So, maybe we'll just have somebody else sign this." I looked at it, said, "Well, this is consistent with what I was thinking, it's just a lot shorter. You're not going to put President Hinckley's name on it, and it's going to be a whole lot less controversy if I just continue and accept this and put it forward." So, that's what happened.[25]

Hinckley was given all the versions that were being considered: the original four-thousand-word Evenson article; his twenty-five-hundred-word rewrite with corrections by Reynolds, Ludlow, and Matthews; Evenson's final rewrite that attempts to address Matthews's complaints; and Matthews's rewrite. As Willam Evenson stated above, the final version was penned by President Hinckley with Evenson agreeing to put his name on it to smooth things on their way.

B. Ken Harrison, a BYU physicist who favored evolutionary explanations for life on Earth, sent a letter to Hinckley because he was worried that the article would be too harsh on evolution. In a wry letter, Hinckley responded, "You wrote to me under date of 18 April concerning the proposed article on evolution in the new *Encyclopedia of Mormonism*. I think you will be satisfied with the article when you see it. Every good wish to you. Sincerely, Gordon B. Hinckley."[26]

After Hinckley had spoken, the debate was over.

This controversy over the issue of evolution in creating the *Encyclopedia of Mormonism* bled into another controversy on what information students should be given to learn about evolution at the church schools like BYU.

The BYU Evolution Packet and *Mormonism and Evolution: The Authoritative LDS Statements*

By the early 1990s, things were starting to look up for the acceptance of evolution in the LDS community. The detractors were outgunned by evolution's growing evidence and relevance. It was the reigning paradigm for all of the life sciences, from the microscopic scales informing DNA studies to landscape-level ecological relationships among populations of organisms over large spatial scales. Among LDS biologists and geologists (and other disciplines that used evolution to understand the world in such diverse fields as engineering and literature), organic evolution shifted from being viewed with suspicion and dismissal to an embrace promoting acceptance and accommodation. However, anti-evolution sentiments were still widespread in the church, especially because they were promoted through the many church manuals designed to educate young people in the CES and from the pulpit by leading General Authorities. The work shifted from defense to offense in order to educate the Saints.[27]

Nevertheless, at BYU, student questions about the relationship of the church and organic evolution kept appearing. Evolution was often mentioned in diverse classes, including those taught in biology, geology, archaeology, philosophy, and religion. For his own and other interested students, Jeffery created a collection of papers that spanned many of the differing opinions about evolution found among the General Authorities, religion professors, and others who had commented on that relationship. Jeffery collected these papers initially for students, and the packet contained the complete gamut of ways that leaders and other influencers interacted with the church and the science of evolution. But no one was happy with it because it was more a hodgepodge of various opinions without any organizing principles.

Evenson and Jeffery decided to set to work creating what would become the official BYU Evolution Packet, to which the entry from the *Encyclopedia of Mormonism* was added. Evenson said of that packet that it was to replace the collection that Jeffery had made.[28] The official packet is still used today and, as of this writing, is available on the biology department's website and still influences students.[29]

Jeffery and Evenson were still not happy that there was no clear collection of all of the statements that General Authorities had made on the subject of evolution and decided to create one. Thus, *Mormonism and Evolution: The Authoritative LDS Statements*, offering a complete list of authoritative statements on evolution under the imprimatur of the church, was born. But they thought it important to also note in the books some of the contradictions they found among the General Authorities they were quoting:

> These are the statements that can be relied upon in the face of diverse unofficial views from Church leaders and others, views that may, in fact, be contradictory. Such contradictions, however, are no more than indications of different opinions on issues not fully settled by authoritative statements or modern revelation. Furthermore, all thoughtful, faithful Latter-day Saints share the responsibility to find their own insights on those questions that are not settled by formal statements of the first presidency.[30]

Utah State Legislature's Attempt to Get Intelligent Design Taught in Secondary Schools

Beyond the borders of BYU, there was still much work to be done to convince the Saints that evolution could be religiously benign and be embraced as part of normal scientific practice. Because of this commitment to his personal integrity, Jeffery felt it was disingenuous to continue to not challenge what had inadvisably crept into the Church Education System. CES manuals had fortified the widespread impression, championed by Joseph Fielding Smith, that evolution was satanic and inappropriate to students' understanding of the gospel. His ideas promulgated the notion that this was the only interpretation of LDS beliefs despite the long-documented history of disagreements among General Authorities of the church.[31] This misperception advanced by Fielding Smith was widespread enough that it entered Utah's political landscape through the advocacy of the Discovery Institute, an organization devoted to the promotion of "intelligent design creationism." Intelligent design creationism (often shortened to ID) was designed to mimic a scientific approach to arguments that there is a designer for the universe.

While its ideas were in circulation for much of the late twentieth century, it found its more clear and detailed articulation in Michael Behe's book *Darwin's Black Box.*[32] Behe looks superficially at arguments against

Darwin's view of evolution but focuses on what he thinks are intractable, irrefutable problems that evolution cannot explain. His examples have been refuted in articles in peer-reviewed scientific journals.[33] Behe's book became a touchstone for evangelical efforts to challenge the teaching of evolution in schools, and school boards all over the nation were enjoined to enter the fray and demand that ID be taught side by side with evolution. This was challenged by evolutionary biologists without exception, including theistic evolutionists.

In Dover, Pennsylvania, the school board decided to require a statement that claimed that there were alternative explanations to the processes giving rise to life on Earth and that ID was one of these and should not be excluded from high school curriculums. This was challenged by parents concerned that science classes should be about science and ID was about religion. They sued the school board, and the case *Kitzmiller v. Dover Area School District* was heard between September 26 and November 4, 2005. The case was watched closely by those who believed in evolution and its scientific claims and those who dismissed it on religious grounds, led largely by Protestant Fundamentalist forces. In the end, the judge, a conservative evangelical Christian, wrote as follows:

> The facts of this case make it abundantly clear that the Board's ID Policy violates the Establishment Clause. In making this determination, we have addressed the seminal question of whether ID is science. We have concluded that it is not, and moreover that ID cannot uncouple itself from its creationist, and thus religious, antecedents.[34]

The judge then spoke about those who brought the case to trial:

> The citizens of the Dover area were poorly served by the members of the board who voted for the ID policy. It is ironic that several of these individuals, who so staunchly and proudly outed their religious convictions in public, would time and again lie to cover their tracks and disguise the real purpose behind the ID Policy.

This dishonesty was on display in Behe's visit to BYU in October 2009.[35]

In 2005, the Discovery Institute, an advocate of "ID" creationism, began an attempt to influence school boards nationwide to replace scientific evidence-based biology curriculums that taught evolution with an anti-science fundamentalist Christian creationist doctrine.

A few years earlier, in 2006, the same year as the Dover trial, Utah state legislator Chris Buttars (R-West Jordan) tried to get the Utah state school board to embrace ID creationism to counter discussions of human evolution in the state's classrooms.[36] The board had decided to propose a position statement on evolution to provide students and parents with clear information on how evolution would be handled in its biology curriculum. The proposed statement reflected the current position of evolution in the scientific community, stating that "evolution is accepted as a principle by the scientific community, it should be taught in science classes, it's got credible evidence behind it, but questions still related to it <evolution>, [and] religion and other methods of knowing . . . should be studied elsewhere." Buttars, however, wanted to "insert language that included a statement that humans did not evolve from other species," and that contrary to scientific consensus, "there is not generally accepted agreement in the scientific community or [evidence] that has stood up to scientific scrutiny regarding the evolution of man from any other species." When the school board refused to bow to Buttars's request, he threatened to make it a state referendum and leave it to the voice of the people.[37]

By December 2005, Buttars had requested a protected bill file that was not subject to public scrutiny. While not mentioning ID in the bill, he had been publicly vocal about supporting the agenda of ID and attempted to make claims that humans had not evolved from apelike species and that there were other theories of human origin that should be allowed to be taught in school. The bill was leaked, and its contents became public. It began,

> In order to encourage students to critically analyze theories regarding the origins of life or the origins or present state of the human race, consider opposing viewpoints, and to form their own opinions, the Legislature desires to avoid the perception that all scientists agree on any one theory or that the state endorses one theory over another.[38]

The state curriculum director, Brett Moulding, pushed back, saying that there were "many, many theories about the origin of life that have no scientific basis." Buttars's choice to hide his connection and commitment to ID but publicly support the idea challenged the constitutionality of the bill.

A month after the bill was leaked, it came up for debate in the legislature as SB96. The *Deseret News* published an article discussing the ways different religions handle human origins.[39] It referenced the Evenson and Jeffery authoritative statements book,[40] with Jeffery quoted as saying, "There has been a belief, for years and years and years, that Mormonism and evolution are diametrically opposed," and that he hoped the book would clear up this misimpression. Jeffery also stated that he was a "theistic evolutionist." Evenson, also interviewed, added, "The LDS Church has really been careful over the years not to get into a box where they are taking a position that later gets undermined by science and other developments of human knowledge. I don't think that's well understood by people who would like to have this be a settled issue." The article does a nice job of discussing the Evolution Packet at BYU and the *Encyclopedia of Mormonism*, then quotes President Hinckley as saying, "What the church requires is only belief that Adam was the first man of what we would call the human race. Scientists can speculate on the rest."

On February 8, 2006, the Utah State Legislature debated the merits of a modification of the Buttars bill put forward by Rep. James A. Ferrin (R) because Chris Buttars was in the hospital. He stated that not all scientists agreed on the theory of evolution and that it was not "indisputably proven." Duane Jeffery was the only doctoral-level scientist to appear due to time constraints. He had only a few minutes to respond but offered two points for the body to consider: "I am concerned first of all with the bill, about what do we mean by 'indisputably proven,' and who on earth is going to be the judge of that? Let me give one quick example. I think most people here would accept the idea that HIV as a virus is what causes AIDS. Peter Duesberg at the University of California, Berkeley, denies that fact and the nation of South Africa with their president, Thabo Mbeki, has seized on that, and said, 'Okay, we're not going to treat HIV because Peter Duesberg says it doesn't cause AIDS.' That has cost hundreds of thousands of lives in South Africa. So, when we say, 'indisputably proven,' who's going to be the judge of that? I see nothing here that even sets up any kind of mechanism for that."[41]

He then used a common highway plant, *Atriplex robusta*, to show that species change. He explained that the plant doubled its chromosomes while remaining fertile, even though because of the chromosomal change it could not breed with a plant that matched its parents' DNA, refuting the idea that species cannot evolve.

He then struck at what he sensed might be a chief concern for the state's economic future: "My last point would have to do with damage to the state and our attempt to try to improve the image of the state of Utah and recruiting top-level scientists here. Believe me, this bill will not be seen as indicative that their children can get high-quality education if they come to the state of Utah." Despite these efforts by Jeffery, the bill passed with a vote of 7–6. However, on February 27, 2006, when the bill went to the State House, it was killed in a vote of 46–28 against, due to the efforts of Minority Whip Stephen Urquhart.[42] Jeffery was important in this defeat, especially because his BYU credentials gave him some influence among many members of the church. As the institution's flagship of higher learning, it carried some cachet among many members of the state legislature who were predominately members of the sponsoring church. Jeffery was well known in the Utah community because his widely read column in the *Provo Daily Herald*, "Science and Society," was an important source of accurate science information.

Science and Society: Jeffery Teaches the Broader Community of the Saints Why Science Matters

Jeffery demonstrated a strong commitment to building and strengthening his community. One of the most important elements of Jeffery's life's work to teach the Saints science was a long-running column about science in the newspaper *Provo Daily Herald*. The column, called "Science and Society," covered a rich array of scientific topics for over twenty-five years, from 1990 to 2015. In a serious but friendly style, he enjoined the citizens of Utah Valley to understand the value and importance of science, spanning topics from evolutionary biology, genetics, medicine, conservation biology, climate change, antibiotic resistance, pollution, and scientifically informed insights into agriculture, including the use of transgenic crops. The range and depth of the topics was striking.

It became a popular column that had tremendous influence. Steve Leavitt, one of the current faculty in BYU's biology department, recalls visiting his grandmother as an early teen and rifling through her stacks of saved newspapers to read Jeffery's column, instilling a curiosity for exploring the world through science.[43]

Jeffery's first column appeared in the *Provo Daily Herald* on March 12, 1990, and tackled a pervasive myth: "Scientists say . . . " as if science speaks with one voice, when in reality, science seeks consensus among many scientists. That he starts with dispelling myths is telling in Jeffery's work because he would spend much of his life teaching students that science is one of the most important tools we have. His general categories were broad, including understanding science, the relationship of science to religion, and even problems with science. Throughout the years the column ran, Jeffery did not retreat from controversy. For example, he covered the uncertainties of the pesticide dioxin (May 27, 1991); lead poisoning (November 18, 1991); race and IQ (March 6, 1995); wilderness areas and designation (May 20, 1996); dyslexia (November 18, 1996); genes and genealogy (February 24, 1997); economic and ecological resources (June 23, 1997); new hope for tuberculosis treatment (February 9, 2005); the value of biodiversity (April 21, 2010); human DNA—various origins? (June 9, 2010); amphibians in worldwide decline (January 25, 2012); roundup of recent climate data (November 28, 2012); our world without evolution (April 3, 2013); an update on vaccination progress (March 26, 2014); a war on science? (March 12, 2015); and many more topics of interest to the Utah community where he lived.

It was also not without controversy. For example, in 1995, he did a two-part series on new data on IQ and race, which discredited the idea that IQ was racially based and argued for environmental factors being the most significant driving factor. The comments were blistering. In an article about his column in the *Provo Daily Herald* describing his decision to end the series on race and IQ, he included "not wanting to risk readers screaming censorship for not having their racist feedback published." He added with disgust about the comments that had flooded into the editorial office of the paper: "It was just horrible, I thought I was reading something out of the South in the 1850s."[44]

When Jeffery began his column in 1990, newspaper circulation in the United States was the highest it had been since the 1940s, with over 62 million subscribers.[45] If these trends were representative of Utah County (and there is little reason to doubt the readership was proportional to national rates), Jeffery was reaching over half of the households in the county. So, a significant number of the households had access to Jeffery's science writing. However, 1990 also marks the time when newspaper circulation began a precipitous decline due to cable news, internet sources, and other emerging forms of media and discourse, one that continues to this day.

In 2014, due to the need to care for a sibling, he decided to make the column monthly instead of weekly—for twenty-four years, he had written an article every week.[46] There were missing weeks due to various circumstances, but a conservative estimate means that he wrote over 1,100 articles in those years.[47] His column ended in 2015, shortly after his retirement from BYU. Those monthly articles that followed that last year would be his last. However, his contributions to science education among the Saints are monumental. His work garnered a number of well-deserved awards and recognition.

The Value of Jeffery's Work

On March 16, 2018, at the annual meeting of the National Science Teaching Association held in Atlanta, Georgia, Duane Jeffery was given the National Science Teaching Association's Presidential Citation.[48] The award recognizes his teaching, his work on the board of directors (1994–2009) for the National Center for Science Education, and for sitting on its Advisory Council. The announcement for this prestigious award states, "The award is conferred on a person or organization that has significantly promoted science education through extraordinary contributions."

In addition to this national award, Jeffery received many awards within the institution, including the Associated Students of BYU Master Teacher Award, 1971; BYU Honors Professor of the year, 1979; the College of Biology and Agriculture Distinguished Teacher Award, 1984; the Alcuin Award, 1986–1991; the Karl G. Maeser Distinguished Teacher Award, 1991; the Department of Zoology Outstanding Service Award, 1996; and the College of Biology and Agriculture Distinguished Teacher Award, 2004.

It is hard to overestimate the value of Jeffery's contributions to bringing evolution into the front of religious education in evolutionary science. He has had a profound influence on many, and these awards are just a shadow of that influence.

A Personal Reflection by the Author on the Current State of Evolution Education at BYU

At the fifty-year celebration of evolution being taught at BYU in 2023, Clayton White and Duane Jeffery's contributions were recognized, feted, and honored. They received an award from the Monte L. Bean Museum, and its director, Professor Mike Whiting, traced ways that BYU had become a nationally recognized and well-respected home of evolutionary studies.

One of the drivers of evolution education at BYU has been hiring faculty from top academic universities who have actively obtained external funding to study evolution. Whiting notes BYU was one of the first universities in the world to purchase and establish a DNA Sequencing Center, and this has given it front-row access to some of the most significant developments in evolutionary theory, including elucidating Earth's tree of life, constructed using DNA to map relatedness of all the species, giving a family tree for how Earth's organisms have evolved from life's establishment over 3.5 billion years ago. This allowed BYU faculty to obtain four grants in the highly competitive National Science Foundation (NSF) program, Assembling the Tree of Life. This has led to over 112 federally funded NSF grants, with over $20,500,000 coming to the university to support this work. In addition, BYU biology faculty have obtained over $3.5 million from other federal and other funding sources to secure over forty successful grants.

In addition to the funding success of many of the life science faculty, BYU Bean Life Science Museum curates over five million biological specimens used by academic institutions and research programs from all over the world for academic study and research, including most of the major taxa of life on our planet such as lichens, plants, microfauna like nematodes, insects, fish, reptiles and amphibians, birds, and mammals. Moreover, the museum provides programs to help educate the public on the life sciences from an evolutionary perspective. Currently on display at the museum (and in the Life Science Building at BYU) is the museum's first display of human evolution, including re-creations of the fossil evidence of the hominims (direct ancestors of modern humans), and other associated hominid dead ends that form the leaves of the branches on the lineage that eventually led to the human species, *Homo sapiens*. The museum has also recently added an exhibit on the "Tree of Life," the genetic construction of which involved a number of BYU life science professors.

We also see in the church a willingness to openly discuss evolution, such as in the *Gospel Topic Essays* on evolution and in the accessible overview of some of evolution's history in the *Saints* series published by the church. Jamie Jensen, in the biology department at BYU, and her collaborators from many institutions are actively researching how institutions with a commitment to both science and faith navigate and instantiate a dynamic reconciliation between these two vital ways of understanding the world from different perspectives and proclivities. She is the president-elect of the National Society for the Advancement of Biology Education Research. She is a member of the Broader Social Impacts Committee of the Smithsonian's Human Origins Project. Jensen, senior author, and first author William Bradshaw looked at how evolution has changed based on surveys they conducted from two cohorts of students: one conducted by Bradshaw (1986–96) and a second by Jensen's lab (2014–16). The findings revealed some needed encouragement for those interested in reconciling evolution and the church's teachings:

> We report here a significant shift toward acceptance [of evolution], compared to attitudes 30 years earlier, by students at Brigham Young University, which is owned and operated by the LDS church. This change appears to have multiple explanations. Students currently entering the university have been exposed to a much-improved introduction to evolution during high school. More importantly, there has been a significant decrease in negative messaging from Church authorities and in its religious education system.[49]

Despite these positive steps, the battle for which Jeffery and many others have fought so hard to reconcile and create the perception that the science of evolution and faith are compatible is still held in suspicion by many members of the church, likely because of evolution's widespread dismissal in CES and by the well-known opinions by former and influential leaders of the church as noted in the previous chapters.[50] In 2014 (the latest data from the Pew Foundation but close to the study conducted by Bradshaw et al. above), only 11 percent of Mormons said humans evolved by natural processes, and 30 percent answered humans evolved due to God's design. And 51 percent answered that humans have "always existed in the present form." Jana Riess and Benjamin Knoll found much the same thing in the research they conducted in the 2016 Next Mormons Survey with the question, "God created Adam and Eve sometime in the last 10,000 years, and humans did not evolve from other forms of life." Sixty-three percent of men and 73 percent of women answered that this was true.[51]

Jeffery's contributions continue to resound through the years. Many active research biologists and laypeople have benefited from and been influenced by his commitment to science, evolution, and truth. He has always been exemplary in his ethical commitment to the idea that science is the best way to learn about the truth of the universe and that there are no incompatibilities between his scientific commitments and his lifelong desire to understand faith. His legacy allowed BYU to become one of the most active and developed research programs in evolution of any religiously affiliated university in the world. The influence of this farm boy from Delta, Utah, turned out to have affected the LDS people and the science of evolution in profound ways that honor everything BYU stands for and from the beginning hoped to be.

In my last interview with Jeffery, I wanted to give him the last word, so I asked him as we wrapped up, “What insight would you like to impart in the world?” His answer is a fitting place to close this book:

> I learned at an early age how powerful faith can be. In a person’s life and in a person’s family. Faith was incredibly important in my family. I’m very devout. But then the day came when I could see that faith was going too far, that people were losing their ability to think about the world around them. . . . Many people of deep faith are not people who are reliable when it comes to what’s going on in the world around them, even in matters that rely on their faith. In fact, their faith can cover up a whole lot of stuff. . . . I think I’ve tried to come to recognize the strength and the limitations of faith. I’ve tried to recognize the incredible strength of the scientific method. I realize that it can be taken a little far, too, when some people try to apply it to places where it can’t be applied. It doesn’t apply to everything in this world. In a definitive sense, you can speculate on what applying a rational method to some of these things means, but you cannot insist that you’ve got it all right. The questions are too big yet for that. So that’s about where I sit. I guess [people will] say, “I don’t know what you believe!” Maybe, but I’m happy with what I don’t believe—because I know what I believe and what I don’t believe.

Bibliographic Essay

Duane Jeffery's impact and influence are reflected in the breadth and depth of his commitment to students and to the idea that truth matters more than interpretive frameworks in which that truth is often shaped and contextualized for human understanding. Jeffery's scientific contributions, while significant, I will leave to the side briefly to focus here on these commitments and on his efforts as a public figure in communicating evolution to the public. His scientific productivity compared to today is small but was written in a day of much trial and error in which fruit flies had to be reared and processed by hand with poor methods for extracting and analyzing DNA. Compared to Berkeley its research funding was meager.

Jeffery's papers are found in Special Collections, J. Willard Marriott Library at the University of Utah. The collection name is "Duane E. Jeffery papers, 1814–2011"; Collection Number ACCN1372. The site contains the following summary: The "papers consist of professional and administrative materials, research materials, subject files, and professional writings by Jeffery. Jeffery was a distinguished professor at Brigham Young University (BYU) specializing in evolutionary genetics." The collection is extensive, consisting of 225 boxes of material including correspondence with students, general authorities, university administrators, scientists, educators from many universities and other organizations; writings and drafts of his published writings; extensive files on many people and subjects. Currently, the archive can be found at https://archiveswest.orbiscascade.org/ark:/80444/xv27096.

Interviews

This project included creating raw digital sound files and transcripts of several interviews, prepared by my research assistant, Beck Seamons. The interviews include a 2009 interview (comprising five separate sound files, ~2 h 48 m), which I conducted with Rebecca Sperry, one of Jeffery's students, about his life. For the purposes of this book and to clarify things about his experiences and life, I conducted two interviews with Duane Jeffery and his wife, Kaye (née Westover) Jeffery, on August 16, 2022 (~1.5 hours) and another in December 2022, which included Kaye Jeffery, Kodi Jeffery, and student research assistants Beck Seamons and Ethan Adams (~2 hours). Additional interviews now in the Jeffery Collection at the University of Utah include William Bradshaw, former BYU professor and colleague of Jeffery, conducted on January 20, 2023 (~1 hour); William Evenson, former BYU professor and academic vice president (~1 hour); Clayton White, former BYU professor and colleague of Jeffery (~2 hours).

Gary Bergera interviewed Jeffery in 1981. He sent me a transcript of the interview to use for this book.

Writings About Evolution and Related Issues Specifically for Latter-day Saint Audiences

Duane Jeffery was ever active in trying to communicate his thought to the Saints directly. He cared very much about the truth. During his years in training and throughout his years as a scientist, he was concerned that science is not co-opted by ideological claims, whether from religion, apologetic efforts, or politics. Indeed, he argued, any source that tried to make claims about the natural world without reference to science should immediately be held in suspicion if it disagreed with science or lacked relevant support from that source. To reach the Saints, he published some articles in Latter-day Saint–relevant sources like *Dialogue* and *Sunstone*, at the time more controversial journals than they are currently, but also in the peer-reviewed scientific literature.

Evenson, William E., and Duane E. Jeffery. *Mormonism and Evolution: The Authoritative LDS Statements*. Greg Kofford Books, 2005. Equal authorship.

Evenson, William E., and Duane E. Jeffery. "BYU Evolution Packet." BYU, 2009, accessed December 18, 2024, https://biology.byu.edu/https:/brightspotcdn

.byu.edu/od/cc/3094a8654f60b552fa97c3df9999/evolution-packet.pdf. Packet given to BYU students to highlight the relationship of the church to evolution. Equal authorship.

Jeffery, Duane E. "Intersexes in Humans: An Introductory Exploration." *Dialogue: A Journal of Mormon Thought* 12, no. 3 (1979): 107–13.

Jeffery, Duane E. "Noah's Flood: Modern Scholarship and Mormon Traditions." *Sunstone*, October 2004, 27–45.

Jeffery, Duane E. "Review of Reed C. Durham, Jr. and Steven H. Heath's *Succession in the Church.*" *BYU Studies* 11, no. 2 (1971): 207–10.

Jeffery, Duane E. "We Don't Know: A Survey of Mormon Responses to Evolutionary Biology." In *Science and Religion: Toward a More Useful Dialogue*, edited by Wilford M. Hess, Raymond T. Matheny, and Donlu D. Thayer. Geneva, IL: Paladin House, 1979.

Norman, Keith E., and Duane E. Jeffery. "Thoughts on Mormonism, Evolution, and Brigham Young University." *Dialogue: A Journal of Mormon Thought* 35, no. 4 (2002): 1–18.

Stephens, T. D., D. J. Meldrum, F. B. Peterson, and Duane E. Jeffery (Foreword). *Evolution and Mormonism: A Quest for Understanding*. Signature Books, 2001.

Stutz, H. C., and Duane E. Jeffery (Foreword). *"Let the Earth Bring Forth": Evolution and Scripture*. Greg Kofford Books, 2011.

Publications for Wider Audiences

Jeffery, Duane E. "Dealing with Creationism." *Evolution* 37, no. 5 (1983): 1097–1100.

Jeffery, Duane E. "Review of Spencer Wells, *Deep Ancestry: Inside the Genographic Project.*" *Western North American Naturalist* 68, no. 2 (2008): 260–61.

Jeffery, Duane E. "Reviews of Three Books on Human Sexuality and Gender." *Politics and the Life Sciences* 23, no. 2 (2005): 71–74.[1]

"Science and Society": A Twenty-Five-Year Column in the *Provo Daily Herald*

Jeffery's most sustained effort in supporting his community's efforts to understand science centers on his "Science and Society" column.

While the *Provo Herald* did not reach a national audience, it was widely read by Jeffery's target audience, the Latter-day Saints community, one he recognized was badly in need of further science literacy. His background as a member of the church who had been engaged with science and its history for the better part of his life uniquely positioned him to offer useful readings not only of science's past but also an accurate contextual engagement with its current trends and directions, and with issues of most concern to his times and people. Here is a sampling of topics, in no particular order: evolution, how new species are formed, science and religion, creationism, intelligent design, genetics and the history of genetics, current trends in medical science, pseudoscience, Noah's flood, science education in the United States, antibiotic and insecticide resistance, transgenic crops, scientism, religious experience, cloning, cross-species organ transplantation, geoengineering, carbon dating, human activity as a selective force in species change, the Pennsylvania Dover trial on teaching intelligent design in the classroom, disease processes like mad cow disease, linguistics and the origin of languages, ecology and conservation, ethics, the biology of homosexuality, climate change, recent discoveries

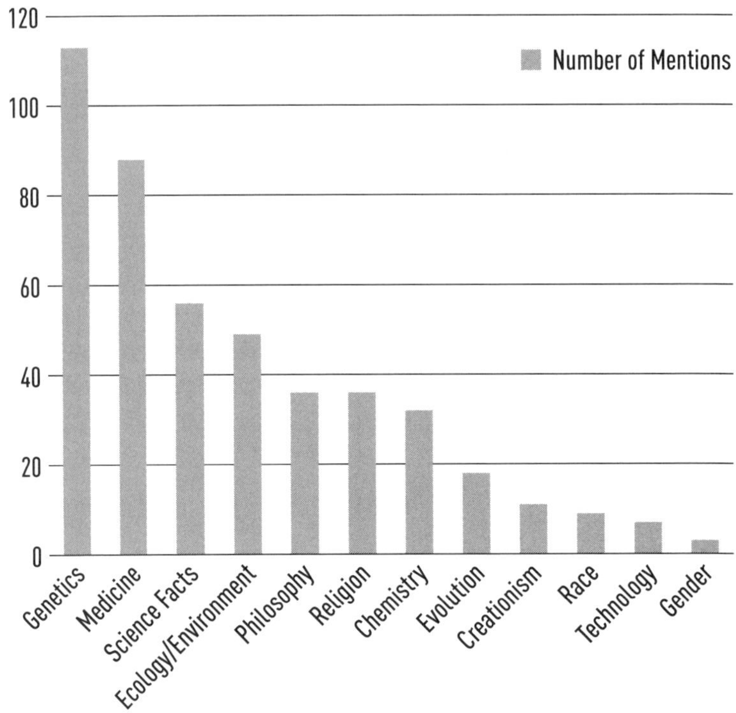

Topics addressed in Jeffery's "Science and Society" column.

in human evolution, overpopulation, water and its importance in the West, genetic mutations and human health, and the list goes on. This is just a small fraction of the more than eleven hundred articles he wrote during the column's run. We located four hundred entries of the column and took them as a random sample. Breaking down the topics into twelve broad categories, this table gives a quantitative sense of the distribution of these topics.

His column has a homey, friendly style while being direct and to the point, evincing a determined stance that both scientific findings and conversations about science need to be taken seriously, especially regarding religion. While his articles are always forceful, he makes it clear he craves conversation about these subjects. His approach is reasoned, factual, and well contextualized, situating his explorations of science in things that matter to people's lives and cultural context. One example of this is found in an entry for his column from June 4, 1998, in which he writes of a meeting he attended of the American Association for the Advancement of Science to begin conversations about science and religion.[2] Jeffery notes both scientists and theologians attended. He talks about the tone of the meeting and its attempts at conversations. It was not meant to be a free-for-all of raucous opinions in a metaphoric shouting match among differing views. Jeffery notes, "It was recognized by all present that society can no longer afford the schism of the past centuries. But it was also clear that synthesis and rapport will not gain discussants a place at the table merely by wishing to be there or by blustering. One had better be prepared to address matters with information and honesty." In a conversation between a couple of people sitting behind him—a geologist and a creationist—the latter asked, "I wonder how many of these theologians are creationists?" Jeffery notes that the geologist is a bit uncomfortable but offers, "Well . . . if you are using that term in the popular sense, I doubt that any of them are." The creationist asks, "Why not?" and the geologist answers, "Well . . . because they have not yet learned to be honest with well-established data." At which the conversation stops. Jeffery believes that man left, and he did not see the creationist again during the rest of the meeting. He laments, "I regretted his absence. Productive communication will require us all to listen to the 'others.'"

This exchange suggests how willing Jeffery was to let people share their perspectives. He honored genuine conversation. Of course, he had a stake in such a conversation, and as in the case described earlier with Duane Gish, he demanded that whoever he was engaging use reason, well-attested facts, and a willingness to calmly debate the merits of differing views.

Jeffery followed the science of evolution constantly through his column, which spanned a period that included new findings he was reading in the primary literature, the Dover trial, religious views on the creation, as well as how evolution has impacted agriculture and other human endeavors. For example, Jeffery used his column to demonstrate the ways evolution affected everything from the practice of medicine to what ends up in our kitchen cabinets, driving home the point that it was not just about dinosaurs, fossils, and old bones. Moreover, he offered nuanced interactions between evolution and Earth's organisms, even the smallest living things on the planet, like his discussions on bacteria. He addressed several dimensions of their interactions with humans. He looked at resistance in pathogens like bacteria and explained how they develop resistance through a combination of selection pressure and the overuse of certain drugs.[3] In another view of the human evolution, he explored how the ability to digest lactose and lactose intolerance is affected by our gut bacteria (microbiome)[4] and in another entry writes about how those microbiomes are really a necessary part of our body without which we would not survive.[5] He explored evolution and bacteria, writing about how researchers introduced a bacterial species of *Wolbachia* into mosquitoes to control the spread of malaria by disrupting their ability to reproduce.[6] In his thinking about bacteria and humans, as in many other settings, it is clear he is never standing still but follows several important scientific developments in evolution wherever they might lead and whatever biological or biologically adjacent field might provide insight into exciting findings that help us understand our place in the universe.

In a final example, I want to explore how he writes about ecological issues, including ethical responses to ongoing worldwide ecological crises like climate change. In his column published on April 21, 2010,[7] the day before Earth Day,[8] Jeffery writes on the value of biodiversity. He starts with a story about a student who asks an unidentified guest speaker why the Death Valley pup fish should be protected and preserved, then asks, "What good is it?" The speaker thinks and then asks, "What good are you?" Jeffery then lays out a plan for protecting Earth's biodiversity and for conservation with four elements: (1) a utilitarian approach that looks at the economic impact of the plants and animals required for human sustenance; (2) an approach that looks at indirect values of such services as soil binding, carbon sequestering by plants, as well as other ecosystems services; (3) an approach that acknowledges aesthetic values one derives from being in nature for recreation and contemplation; and (4) the intrinsic value that all creatures have as part of the creation (intrinsic value was what Jeffery marked as the answer to the speaker's question about what good the student was).

Jeffery's exploration of ecology, species biodiversity was a theme he visited frequently, as with his exploration into the effects of anthropogenic climate change, the theme of his June 18, 2008, article on how oceans are threatened by human activities.[9] In this entry, he looks at three articles from the scientific literature, the first from the journal *Nature*, which looks at President Anote Tong of the island nation of Kiribati, whose land is submerging due to rising sea levels caused by warming oceans. He then looks at coastal flooding happening for the same reason in Bangladesh. Last, he looks at Pacific gyres made of millions of tons of plastic garbage, collecting from widespread oceanic dumping of human trash into what was once viewed as the unlimited vastness of these waters.

In all the articles published during the twenty-five years he wrote the column, he brought science home to Utah's population, including a large number of church members. Like a local version of evolutionary biologist Stephen Jay Gould, whose long-running *Natural History* magazine essays changed the face of public discourse about science, Jeffery's influence on the Saints through his articles was a profound effort in science education.

Notes

Chapter One. Biological Sketch of a Unique Ape

1. Duane E. Jeffery, "Duane and Kaye Jeffery Interview August 2022," by Steven L. Peck and Beck Seamons, 2022, 30. Transcript ln 16–96.

2. Duane E. Jeffery, "Duane E. Jeffery 2009 Interview," by Steven Peck and Rebecca Sperry, transcription by Beck Seamons, December 4, 2009, 46. Transcript ln 223–54.

3. Jeffery, "Duane E. Jeffery 2009 Interview," transcript ln 60–69.

4. Kahlile Mehr, "The Trial of the French Mission," *Dialogue: A Journal of Mormon Thought* 21, no. 3 (1988): 27–45.

5. Jeffery, "Duane E. Jeffery 2009 Interview," transcript ln 125–30.

6. Jeffery, "Duane E. Jeffery 2009 Interview," transcript ln 151–59.

7. Jeffery, "Duane E. Jeffery 2009 Interview," transcript ln 768–826.

8. Especially in light of future interactions with Bruce R. McConkie, *Mormon Doctrine: A Compendium of the Gospel* (Bookcraft, 1958).

9. Jeffery, "Duane E. Jeffery 2009 Interview," transcript ln 89–91.

10. Jeffery, "Duane E. Jeffery 2009 Interview," transcript ln 112–15.

11. Joseph Fielding Smith, *Man, His Origin and Destiny* (Deseret Book, 1954).

12. Jeffery, "Duane E. Jeffery 2009 Interview," transcript ln 197–98.

13. Jeffery, "Duane E. Jeffery 2009 Interview," transcript ln 198–202.

14. Jeffery, "Duane E. Jeffery 2009 Interview," transcript ln 208–13.

15. Nephi Anderson, *Dorian* (Bikuben Publishing, 1921).

16. Henry Drummond, *Natural Law in the Spiritual World* (J. Pott, 1890).

17. Duane E. Jeffery and Keith E. Norman, "Thoughts on Mormonism, Evolution, and Brigham Young University," *Dialogue: A Journal of Mormon Thought* 35, no. 4 (2002): 1–18.

18. Duane E. Jeffery, "Factors Influencing Elk Distribution on Willow Creek Summer Range, Utah" (MS thesis, Utah State University, 1963), https://digitalcommons.usu.edu/etd/8631/.

19. Jeffery, "Duane E. Jeffery 2009 Interview," transcript ln 228–30.
20. Jeffery, "Duane E. Jeffery 2009 Interview," transcript ln 453.
21. Duane E. Jeffery, "Interview with Gary Bergera," by Gary Bergera (1981), 16.
22. Duane E. Jeffery, "Position Effects at the Hairy Locus in Drosophila Melanogaster," *Genetics* 91, no. 1 (1979): 105–25.
23. Jeffery, "Duane E. Jeffery 2009 Interview," transcript ln 245–55.
24. Jeffery, "Duane E. Jeffery 2009 Interview," transcript ln 281–83.
25. Jeffery, "Duane E. Jeffery 2009 Interview," transcript ln 285–90.
26. Jeffery, "Duane E. Jeffery 2009 Interview," transcript ln 481–502.
27. Jeffery, "Duane E. Jeffery 2009 Interview," transcript ln 729–39.
28. Kristine L. Haglund, *Eugene England: A Mormon Liberal. Introductions to Mormon Thought* (University of Illinois Press, 2021), 38.
29. For more on England's committed faith but more naive confrontations with General Authorities, see Terryl Givens, *Stretching the Heavens: The Life of Eugene England and the Crisis of Modern Mormonism* (University of North Carolina Press, 2021), 253–85.
30. Haglund, *Eugene England*, 118.
31. Jeffery, "Duane E. Jeffery 2009 Interview," transcript ln 923–49.

Chapter Two. Origin Stories

1. Duane E. Jeffery, "Duane E. Jeffery 2009 Interview," by Steven Peck and Rebecca Sperry, transcription by Beck Seamons, December 4, 2009, 46. Transcript ln 313–30.
2. Jeffery, "Duane E. Jeffery 2009 Interview," transcript ln 364–65.
3. Jeffery, "Duane E. Jeffery 2009 Interview," transcript ln 352–58.
4. Jeffery, "Duane E. Jeffery 2009 Interview," transcript ln 914.
5. B. Waterman and B. Kagel, *The Lord's University: Freedom and Authority at BYU* (Signature Books, 1998), 125–26.
6. Gregory A. Prince and William Robert Wright, *David O. McKay and the Rise of Modern Mormonism* (University of Utah Press, 2005), 179.
7. Prince and Wright, *David O. McKay and the Rise of Modern Mormonism.*
8. Prince and Wright, *David O. McKay and the Rise of Modern Mormonism*, 168–79. The relationship between fundamentalist young-earth creationism and right-wing fears of creeping socialist and communist influences emerged in the 1950s and '60s in the post–World War II era, especially with the publication of the "blockbuster," John C. Whitcomb and Henry M. Morris, *The Genesis Flood: The Biblical Record and Its Scientific Implications* (Presbyterian and Reformed, 1961). This book strongly is influenced by young-earth creationism. This national influence among fundamentalist perspectives likely supported LDS leanings toward young-earth creationism among many LDS leaders. Carl R. Weinberg,

Red Dynamite: Creationism, Culture Wars, and Anticommunism in America (Cornell University Press, 2021), 166–67.

9. Prince and Wright, *David O. McKay and the Rise of Modern Mormonism*, 177.

10. For more details about the history of academic freedom at BYU, see Scott H. Abbott, *Dwelling in the Promised Land as a Stranger: Personal Encounters with Mormon Institutions* (By Common Consent Press, 2021), and Sara M. Patterson, *The September Six and the Struggle for the Soul of Mormonism* (Signature Books, 2023).

11. Clayton White, "Clayton White Interview, with Steven Peck (12/12/22)," by Steven L. Peck, December 12, 2022, 1–21. Transcript ln 102–20.

12. White, "Clayton White Interview," transcript ln 181–88.

13. Jeffery, "Duane E. Jeffery 2009 Interview," transcript ln 499–510.

14. Jeffery, "Duane E. Jeffery 2009 Interview," transcript ln 502–37.

15. White, "Clayton White Interview," transcript ln 157–72.

16. William W. Winder, *History of the Department of Zoology and Entomology and the Department of Zoology at Brigham Young University*, University Archives, L. Tom Perry Special Collections, Harold B. Lee Library, MSS 8972, LTPSC, Brigham Young University, Zoology Department, Brigham Young University, 2016, https://archives.lib.byu.edu/repositories/byu-archives/resources/upb_mss8972, p. 18.

17. Devery S. Anderson, "A History of Dialogue, Part Two: Struggle Toward Maturity, 1971–1982," *Dialogue: A Journal of Mormon Thought* 33, no. 2 (2000): 1.

18. Armand L. Mauss, *The Angel and the Beehive: The Mormon Struggle with Assimilation* (University of Illinois Press, 1994).

19. John Haught, *God After Darwin* (Taylor & Francis, 2021), i.

20. See, for example, John Haught, *Deeper than Darwin: The Prospect for Religion in the Age of Evolution* (Avalon, 2009).

21. Elliott Sober, "Evolution, Population Thinking, and Essentialism," *Philosophy of Science* 47, no. 3 (1980): 350–83), http://www.jstor.org.byu.idm.oclc.org/stable/186950.

22. Erasmus Darwin, *Zoonomia; or the Laws of Organic Life* (Byrne, 1800), 565.

23. Charles Darwin, *On the Origin of Species* (Folio Society, 2006), 49.

24. Theodosius Dobzhansky, "Nothing in Biology Makes Sense Except in the Light of Evolution," *American Biology Teacher* 75, no. 2 (2013): 87–91.

25. One interesting expansion of evolutionary science has been the field of bioinformatics, which has greatly enhanced the church's efforts in genetics and revealing ancestry in human populations.

26. See Benjamin E. Park, "Salvation Through a Tabernacle: Joseph Smith, Parley P. Pratt, and Early Mormon Theologies of Embodiment," *Dialogue: A Journal of Mormon Thought* 43, no. 2 (2010): 1–44.

27. Steve L. Peck, "Trajectories in the Evolution of Mormon Studies on Faith and Science," *Mormon Studies Review* 6, no. 1 (2019): 67–90.

28. Brigham Young, "Remarks by President Brigham Young, Delivered in the Tabernacle, Salt Lake City, May 14, 1871," *Journal of Discourses Online*, http://www.josephsmithfoundation.org/journalofdiscourses/speakers/brigham-young/attending-meetings-religion-science-geology-the-creation/?highlight=science.

29. James E. Talmage, *First Book of Nature* (Contributor Company, 1888).

30. James E. Talmage, *Domestic Science: A Book for Use in Schools and for General Reading* (George Q. Cannon, 1892).

31. Tim S. Reid, "Mormons and Evolution: A History of B. H. Roberts and His Attempt to Reconcile Science and Religion" (PhD diss., Oregon State University, 1997), 75.

32. Reid, "Mormons and Evolution," 81.

33. Susan L. Trollinger and William Vance Trollinger Jr., *Righting America at the Creation Museum: Medicine, Science, and Religion in Historical Context* (Johns Hopkins University Press, 2016), 3.

34. George M. Marsden, *Fundamentalism and American Culture* (Oxford University Press, 2006), 119–10.

35. Trollinger and Trollinger, *Righting America at the Creation Museum*, 4.

36. Davis Bitton, "N. L. Nelson and the Mormon Point of View," *Brigham Young University Studies* 13, no. 2 (1973): 157–71.

37. Nel Lars Nelson, *Scientific Aspects of Mormonism, or Religion in Terms of Life* (Putnam's, 1904).

38. Nelson, *Scientific Aspects of Mormonism*, v.

39. Nelson, *Scientific Aspects of Mormonism*, 20.

40. John A. Widtsoe, *Joseph Smith as Scientist: A Contribution to Mormon Philosophy* (General Board, Young Men's Mutual Improvement Associations, 1908).

41. Peck, "Trajectories in the Evolution of Mormon Studies on Faith and Science."

42. Joseph F. Smith and Edward H. Anderson, "Priesthood Quorum Table: Origin of Man," editorial, *Improvement Era* 6 (April 1910): 570.

43. Stephen C. Taysom, *Like a Fiery Meteor: The Life of Joseph F. Smith* (University of Utah Press, 2023), 349–50.

44. Taysom, *Like a Fiery Meteor*, 350.

45. Matthew Bowman, *Joseph Fielding Smith: A Mormon Theologian* (University of Illinois Press, 2024), 71–72.

46. Henry Peterson, "Biographical Note: Henry Peterson Papers, 1895–1940," COLL MSS 105, Special Collections and Archives Department, updated February 6, 2025, https://archiveswest.orbiscascade.org/ark:80444/xv01531#prefercitelD, p. 1.

47. Gary James Bergera and Ronald Priddis, *Brigham Young University: A House of Faith* (Signature Books, 1985), 135.

48. Bergera and Priddis, *Brigham Young University*, 135.

49. Bergera and Priddis, *Brigham Young University*, 136–37.

50. Bergera and Priddis, *Brigham Young University*, 138.

51. Bergera and Priddis, *Brigham Young University*, 31.

52. Ralph V. Chamberlin, *Life and Philosophy of W. H. Chamberlin* (Deseret News Press, 1925), 148–52.

53. Chamberlin, *Life and Philosophy of W. H. Chamberlin*, 211.

54. Chamberlin, *Life and Philosophy of W. H. Chamberlin*, 139. I quote this in full because I could write the same about my experience as a student under Duane Jeffery. He had similar views and structures about student learning that did not fear exposure to perspectives differing from simplistic and shallow views of the LDS faith and trusted our faith could withstand scientific and other academic challenges and perplexities.

55. James M. McLachlan, "W. H. Chamberlin and the Quest for a Mormon Theology," *Dialogue: A Journal of Mormon Thought* 29, no. 4 (1996).

56. Davis Bitton, "A Master Work of Mormon Theology?" in *The Truth, the Way, the Life: An Elementary Treatise on Theology by B. H. Roberts*, ed. John W. Welch (BYU Studies, 1996), 561.

57. Bowman, *Joseph Fielding Smith*, 41–43.

58. Bowman, *Joseph Fielding Smith*, 54–56.

59. Ronald L. Numbers, *The Creationists: From Scientific Creationism to Intelligent Design* (Harvard University Press, 2006), 96–109.

60. Richard Sherlock, "A Turbulent Spectrum: Mormon Reactions to the Darwinist Legacy," *Journal of Mormon History* 5 (1978): 33–59.

61. Sherlock, "A Turbulent Spectrum," 32–33.

62. Richard Sherlock and Jeffery E. Keller, "The B. H. Roberts/Joseph Fielding Smith/James E. Talmage Affair," in *The Search for Harmony: Essays on Science and Mormonism*, ed. Gene A. sessions and Craig J. Oberg (Signature Books, 1993).

63. Sherlock and Keller, "The B. H. Roberts/Joseph Fielding Smith/James E. Talmage Affair," 97.

64. William E. Evenson and Duane E. Jeffery, *Mormonism and Evolution: The Authoritative LDS Statements* (Greg Kofford Books, 2005), 67.

65. It should be recognized that humans originating from a special creation was not an uncommon opinion even among some scientists of the time. Darwin's codiscoverer of evolution, Alfred Russel Wallace, argued that the human mind was too complex to have evolved.

66. Brigham H. Roberts and John W. Welch, *The Truth, the Way, the Life: An Elementary Treatise on Theology* (BYU Studies, 1994).

67. Joseph Fielding Smith, *Man, His Origin and Destiny* (Deseret Book, 1954).

68. The three men discussed were Dr. John William Draper, Dr. Andrew Dickson White, and Dr. Robert Andrew Millikan. J. Fielding Smith, *Man, His Origin and Destiny*, 34.

69. Gary J. Bergera, "Review of Francis M. Gibbons. 'Joseph Fielding Smith: Gospel Scholar, Prophet of God,'" *Journal of Mormon History* 19, no. 2 (1993): 159.

70. Bergera, "Review of Francis M. Gibbons,'" 157–62.

71. Smith, *Man, His Origin and Destiny*, 133.

72. Bruce R. McConkie, *Mormon Doctrine: A Compendium of the Gospel* (Bookcraft, 1958).

73. Bertrand Harrison, "The Relatedness of Living Things," *The Instructor*, July 1965, 272–76.

74. Church of Jesus Christ of Latter-day Saints, "Organic Evolution," accessed March 28, 2023, https://www.churchofjesuschrist.org/study/history/topics/organic-evolution?lang=eng.

75. Mauss, *The Angel and the Beehive*, 96.

76. Mauss, *The Angel and the Beehive*, 96–97.

77. Mauss, *The Angel and the Beehive*, 97.

78. T. Benjamin Spackman, "'The Scientist Is Wrong': Joseph Fielding Smith, George McCready Price, and the Ascendency of Creationist Thought Among Latter-Day Saints in the Twentieth Century" (PhD diss., Claremont Graduate University, 2024), 163–64.

79. See, Kristine L. Haglund, *Eugene England: A Mormon Liberal* (University of Illinois Press, 2021), and Terryl Givens, *Stretching the Heavens: The Life of Eugene England and the Crisis of Modern Mormonism* (University of North Carolina Press, 2021).

Chapter Three. Seers, Savants, and Deeper Conversations

1. Lester E. Bush, "Mormonism's Negro Doctrine: An Historical Overview," *Dialogue: A Journal of Mormon Thought* 8, no. 1 (1973): 11–68.

2. Duane Jeffery, "Interview with Gary Bergera," by Gary Bergera, 1981, 23–24.

3. Duane Jeffery, "Curt Stern Papers, Duane Jeffery Correspondence 1969–1975 Box 13. Berkeley Library," ed. Curt Stern, 19–29. Correspondence, 1969–1975, https://search.amphilsoc.org/collections/view?docId=ead/Mss.Ms.Coll.5-ead.xml;query=;brand=default#controlaccess: American Philosophical Society, July 9, 1973, 1973.

4. Jeffery, "Curt Stern Papers."

5. Duane Jeffery, "Duane E. Jeffery 2009 Interview," pp. 1497–505. The specific creationist literature mentioned by Jeffery could not be identified but is a species of attack common in creationist books and pamphlets.

6. Reynolds, "Interview," ln 438–49.

7. Reynolds, "Interview," ln 500–505.

8. Reynolds, "Interview," ln 959–76.

9. Jeffery, "Duane E. Jeffery 2009 Interview," transcript ln 1578–87.

10. Jeffery, "Duane E. Jeffery 2009 Interview," transcript ln 25–26; italics mine.

11. Duane E. Jeffery, "Seers, Savants, and Evolution: The Uncomfortable Interface," *Dialogue: A Journal of Mormon Thought* 34, no. 1/2 (2001).

12. Jeffery, "Seers, Savants, and Evolution," 45–58.

13. Gary J. Bergera, *Conflict in the Quorum: Orson Pratt, Brigham Young, Joseph Smith* (Signature Books, 2002).

14. Joseph F. Smith and Edward H. Anderson, "Priesthood Quorum Table: Origin of Man," editorial, *Improvement Era* 6 (April 1910): 570.

15. It was challenging to obtain an accurate count of the number of times the paper was cited, as many databases do not provide a way to access this information. Using ProQuest under the guidance of Mike Goates, the life & geological sciences librarian at the Harold B. Lee Library at BYU, we were able to locate over fifty peer-reviewed publications, including ten dissertations from public and private universities. The following references are a list of what we were able to find. This is also a reference to much engagement of the academic thought on evolution and the church. Gene A Sessions and Craig J. Oberg, eds., *The Search for Harmony: Essays on Science and Mormonism* (Signature Books, 1993); Thomas G. Alexander, "'To Maintain Harmony': Adjusting to External and Internal Stress, 1890–1930," *Dialogue: A Journal of Mormon Thought* 15, no. 4 (1982); Thomas G. Alexander, "Between Revivalism and the Social Gospel: The Latter-Day Saint Social Advisory Committee, 1916–1922," *Brigham Young University Studies* 23, no. 1 (1983); Devery S. Anderson, "A History of Dialogue, Part Two: Struggle Toward Maturity, 1971–1982," *Dialogue: A Journal of Mormon Thought* 33, no. 2 (2000); Michael R. Ash, "The Mormon Myth of Evil Evolution," *Dialogue: A Journal of Mormon Thought* 35, no. 4 (2002); David H. Bailey, "Scientific Foundations of Mormon Theology," *Dialogue: A Journal of Mormon Thought* 21 (1988); David H. Bailey, "Science and Mormonism: Past, Present, and Future," *Dialogue: A Journal of Mormon Thought* 29, no. 1 (1996); David H. Bailey, "Mormonism and the New Creationism," *Dialogue: A Journal of Mormon Thought* 35, no. 4 (2002); P. L. Barlow, *Mormons and the Bible: The Place of the Latter-Day Saints in American Religion* (Oxford University Press, 2013); Barry R. Bickmore et al., "Science as Storytelling for Teaching the Nature of Science and the Science-Religion Interface," *Journal of Geoscience Education* 57, no. 3 (2009); William S. Bradshaw et al., "A Longitudinal Study of Attitudes Toward Evolution Among Undergraduates Who Are Members of the Church of Jesus Christ of Latter-Day Saints," *PLOS One* 13, no. 11 (2018); David John Buerger, "The Adam-God Doctrine," *Dialogue: A Journal of Mormon Thought* 15, no. 1 (1982); Noel A. Carmack, "Images of Christ in Latter-Day Saint Visual Culture,

1900–1999," *BYU Studies Quarterly* 39, no. 3 (2000); Paul R. Cazier, "Embracing the Flesh: In Praise of the Natural Man," *Dialogue: A Journal of Mormon Thought* 31, no. 2 (1998); Kent C. Condie, "Premortal Spirits: Implications for Cloning, Abortion, Evolution, and Extinction," *Dialogue: A Journal of Mormon Thought* 39, no. 1 (2006); Richley H. Crapo, "Grass-Roots Deviance from Official Doctrine: A Study of Latter-Day Saint (Mormon) Folk-Beliefs," *Journal for the Scientific Study of Religion* (1987); Dan Erickson, "Mormon Millennialism: The Literalist Legacy and Implications for the Year 2000," *Dialogue: A Journal of Mormon Thought* 30, no. 2 (1997); Clyde R. Forsberg and Phillip Gordon Mackintosh, "When Science Fails," *John Whitmer Historical Association Journal* 41, no. 1 (2021); Clyde R. Forsberg Jr. and Phillip Gordon Mackintosh, "When Science Fails: B. H. Roberts, Polar Lost Israelites, Preadamites, and Other Modern Mormon Legends," *John Whitmer Historical Association Journal* 41, no. 1 (2021); Richard R. Haglund Jr., "Inherit the Wind, Mormon Style," *Dialogue: A Journal of Mormon Thought* 34, no. 3/4 (2001); Steven H. Heath, "The Reconciliation of Faith and Science: Henry Eyring's Achievement," *Dialogue: A Journal of Mormon Thought* 15, no. 3 (1982); Duane E. Jeffery, "Seers, Savants, and Evolution: The Uncomfortable Interface," *Dialogue: A Journal of Mormon Thought* 34, no. 1/2 (2001); Jeffrey E. Keller, "Discussion Continued: The Sequel to the Roberts/Smith/Talmage Affair," *Dialogue: A Journal of Mormon Thought* 15, no. 1 (1982); Kurt W. Laird, *Mormon Rhetoric and the Theory of Organic Evolution* (Brigham Young University, 2008); Craig Livingston, "From Above and Below: The Mormon Embrace of Revolution, 1840–1940" (PhD diss., Temple University, 2002); Thomas Allen McIver, "Creationism: Intellectual Origins, Cultural Context, and Theoretical Diversity" (PhD diss., University of California, Los Angeles, 1989); Dale C. Mouritsen, "A Symbol of New Directions: George Franklin Richards and the Mormon Church, 1861–1950 (Utah)" (PhD diss., Brigham Young University, 1982); Keith E. Norman, "Thoughts on Mormonism, Evolution, and Brigham Young University," *Dialogue: A Journal of Mormon Thought* 35, no. 4 (2002); Erich Robert Paul, "Science: 'Forever Tentative'?" *Dialogue: A Journal of Mormon Thought* 24, no. 2 (1991); David Lamont Paulsen, *Comparative Coherency of Mormon (Finitistic) and Classical Theism* (University of Michigan Press, 1975); Allan Dean Payne, "The Mormon Response to Early Progressive Education, 1892–1920" (PhD diss., University of Utah, 1977); Richard C. Poulsen, *Misbegotten Muses: History and Anti-History* (Peter Lang, 1988); Tim S. Reid, "Mormons and Evolution: A History of B. H. Roberts and His Attempt to Reconcile Science and Religion" (PhD diss., Oregon State University, 1997); Paul C. Richards, "Does Paying Tithing Make You a Voting Shareholder? Brigham Young University's Worldwide Board of Trustees," *Dialogue: A Journal of Mormon Thought* 26, no. 3 (1993); E. Jay Richardson, *The Thin Line: Reason and Revelation in Contemporary Mormonism* (University of Calgary Press, 2008); Morris S. Petersen, "Earth," in *Encyclopedia of Mormonism*

(Macmillan, 1992), https://eom.byu.edu/; T. W. Simpson, *American Universities and the Birth of Modern Mormonism, 1867–1940* (University of North Carolina Press, 2016); T. W. Simpson and Gregory L. Smith, "Often in Error, Seldom in Doubt: Rod Meldrum and Book of Mormon DNA," *FARMS Review* 22, no. 1 (2010); Richard Pearson Smith, "Science: A Part of or Apart from Mormonism?" *Dialogue: A Journal of Mormon Thought* 19, no. 1 (1986); W. V. Smith, *Textual Studies of the Doctrine and Covenants: The Plural Marriage Revelation* (Greg Kofford Books, 2018); Stephen Snow et al., "Seers, Savants, and Evolution: A Continuing Dialogue," *Dialogue: A Journal of Mormon Thought* 9, no. 3 (1974); S. G. Southerton, *Losing a Lost Tribe: Native Americans, DNA, and the Mormon Church* (Signature Books, 2004); Thomas Benjamin Spackman, "'The Scientist Is Wrong': Joseph Fielding Smith, George McCready Price, and the Ascent of Creationist Thought Among Latter-Day Saints in the Twentieth Century" (PhD diss., Claremont Graduate University, 2024); Jeremy Talmage, "All but American: Latter-Day Saints in the Early Twentieth Century" (PhD diss., University of Utah, 2021); Grant Underwood, "Early Mormon Perceptions of Contemporary America: 1830–1846," *Brigham Young University Studies* 26, no. 3 (1986); Grant Underwood, *Science in the American Southwest: A Topical History* (University of Arizona Press, 2002).

16. Ezra Taft Benson, "Jesus Christ—Gifts and Expectations," *BYU Devotional*, BYU Speeches BYU, 1974.

17. Duane E. Jeffery, "Ezra Taft Benson File," 1974 to 1976, Duane E. Jeffery papers, ACCN 1372, Box 154, Folder 34 Special Collections, J. Willard Marriott Library, University of Utah, pp. 1–12.

18. In the Church of Jesus Christ of Latter-day Saints, congregations of fifty to four hundred members are called "wards," and several wards (usually around ten) are called a "stake." So a fourteen-stake fireside is a very large gathering.

19. Ezra Taft Benson, "God's Hand in Our Nation's History," vol. 2023, *BYU Species*, Brigham Young University, BYU speeches, 1976.

20. Jeffery, "Duane E. Jeffery 2009 Interview," transcript ln 568–76.

21. Jeffery, "Ezra Taft Benson File," Duane E. Jeffery Papers, 1–12.

22. Gary James Bergera and Ronald Priddis, *Brigham Young University: A House of Faith* (Signature Books, 1985), 166.

23. Jeffery, "Ezra Taft Benson File," Duane E. Jeffery Papers, 13–14.

24. Jeffery, "Ezra Taft Benson File," Duane E. Jeffery Papers, 15–16.

25. Jeffery, "Ezra Taft Benson File," Duane E. Jeffery Papers, 17.

26. Jeffery, "Ezra Taft Benson File," Duane E. Jeffery Papers, 19–21.

27. Benson, "Jesus Christ—Gifts and Expectations."

28. Benson, "God's Hand in Our Nation's History."

29. See, for example, Elder Uchtdorf's October 2013 LDS General Conference talk in which he admits, "And, to be perfectly frank, there have been times when

members or leaders in the Church have simply made mistakes," a rare admission for church leaders. Dieter F. Uchtdorf, "Come Join with Us," *Ensign*, 2013, https://www.churchofjesuschrist.org/study/ensign/2013/11/saturday-morning-session/come-join-with-us?lang=eng.

30. See, for example, Reid L. Neilson and Scott D. Marianno, "True and Faithful Joseph Fielding Smith as Mormon Historian and Theologian," *BYU Studies Quarterly* 57, no. 1 (2018): 6–64.

31. Joseph Fielding Smith, "Opening Address," Conference Report, October 1970 (The Church of Jesus Christ of Latter-day Saints, 1970), 5.

32. Jeffery, "Interview with Gary Bergera," 43.

33. Jeffery, "Interview with Gary Bergera," 43.

34. Wilford M. Hess, Raymond T. Matheny, and Donlu D. Thayer, *Science and Religion: Toward a More Useful Dialogue.* 2 vols. (Paladin House, 1979).

35. William E. Evenson, "Science and Religion: Toward a More Useful Dialogue, Vols. I and II," *Brigham Young University Studies* 20, no. 3 (1980): 313–18.

36. **A personal reflection by Steven L. Peck**—This event took place about the time I arrived at BYU as an undergraduate student in 1981. I stepped onto campus amid the events I am describing in this chapter, placing me, like the ancient Greek historian Herodotus, in the unusual position of being a chronicler of events in which I was deeply involved. Suddenly, I move from being a third-person explorer of record of these events to being an eyewitness.

I was unaware of much of the background I had sketched above. Still, upon arrival I was quickly made aware that a war existed between the biologists and many of the professors of religion. And at first, I was firmly in the camp of Joseph Fielding Smith's thought—that is, that evolution was a tool of the devil. I believed that the church was against evolution. My seminary teacher in Moab, Utah, made his disbelief in dinosaurs very clear, and much of my mission was spent reading the thoughts of general authorities that were, for the most part, staunch anti-evolutionists. I even recall telling an eager investigator that he could not be baptized until he quit "believing" in evolution.

Upon my arrival at BYU, a friend whom I met in the army told me about the evolution class at BYU, and assuming that it was a class against evolution, I beelined it to the bookstore to discover what books they were using to introduce it in BYU classes—intending to collect more evidence against organic evolution. To my surprise, I found a regular textbook that did not dismiss the concept. I started taking zoology classes and found thoughtful professors who were committed to both the church and the gospel as well as science. In addition, David Attenborough's magisterial PBS series *Life on Earth* and Carl Sagan's *Cosmos* were airing. I gathered with a group of friends to watch the story of life on Earth unfold through the process of evolution. The beauty of this story was unveiled through my coursework in the zoology program, and these masterful television programs struck me as necessary and true to the facts gleaned abundantly in the natural world. I was convinced.

However, being stuck between my science professors and religion professors quarreling about where science fits in our church almost made me a casualty of the fight. I saw so much beauty in my science classes and so much narrow, entrenched dogmatism in my religion classes that if only one could be right, I think I would have bowed to the confusion and either become a fundamentalist-leaning creationist church member or left the church to pursue science. Based on my reading of Fielding Smith, I assumed that such a choice was required, as Fielding Smith argued that a person could not believe in both the gospel and evolution. When I took evolution from explicitly faithful science professors, they opened a way to embrace both, without casting aside a deeper faith than the shallow anti-science stance of many of my religion professors. For example, my Pearl of Great Price professor told me that if I continued in my belief in evolution, I would go to hell. (It seemed a nice Mormon terrestrial kingdom hell was not good enough for the likes of me, and he resorted to the rhetoric of a "fire and brimstone" to point toward my looming postmortal fate.) He also tried refuting evolution by pointing out that there were only cat and dog spirits in the preexistence, and nothing in between was possible. The illogic of such flawed thinking seemed childish and ill-informed in light of the depth and wonder I was exposed to in my science and humanities classes.

I brought this up with Jeffery in his office while taking his evolution class. I mused that the "war" would probably be over when my kids were going to school, expressing my belief that the "other side" was losing so badly in the hearts and minds of the biology students; the battle lines could not be maintained. Jeffery looked thoughtful, then said, "No. This battle will continue into the lives of your grandchildren and great-grandchildren." I understand better what he was going through then, and now, as I near retirement, I think his cynicism was warranted.

Even so, I was glad Duane Jeffery, and other professors like him, were there at this vulnerable time when I was sorting out what I wanted to become and how to integrate my growing interest in faith and science. He became, for me and other students, a guiding light, showing me how to embrace the best thinking in the secular world while developing a mature faith that could weather the debates about life's origins on planet Earth.

However, the mixed messages from my science classes and my mandatory religion classes gave me pause as I found myself forming a very negative impression of the religious component of my education. For example, in one of my religion classes, I was presented with Reid Bankhead's small self-published book, *The Fall of Adam, the Atonement of Christ, and Organic Evolution*. The book provided an assortment of cherry-picked anti-evolution quotes from General Authorities of the church that could be found before 1978, when the book was published. *The*

Fall of Adam, the Atonement of Christ, and Organic Evolution typifies the rhetoric of many of my religion teachers during my undergraduate years at BYU.

Bankhead begins with what can be seen as a direct attack on Jeffery's efforts to get a clear picture of the official, authoritative statements of the General Authorities of the church and the central message of his "Seers and Savants" article: The comment such as "the Church has taken no stand on organic evolution" could be made by someone who has neglected the task of properly defining evolution, on the one hand, and the fall of man and the atonement of Christ, on the other hand.

Bankhead then issues a general condemnation of scientists for their insistence on evolution: "In minimal terms, scientists have defined acceptance of organic evolution as acceptance of the principles of common and natural descent. Many of the General Authorities of The Church of Jesus Christ of Latter-day Saints have defined and explained the circumstances incident to the fall of Adam and the atonement of Christ. The two positions are not compatible with one another" (Bankhead, *The Fall of Adam*, 1). He then not only condemns Darwin but also those who used his work: "Karl Marx considered dedicating his book, *Das Kapital*, to Charles Darwin" (Bankhead, *The Fall of Adam*, 1). Following these claims are a selection of quotes from individuals ranging from Brigham Young to Bruce R. McConkie, including some, such as John Widtsoe and James Talmage, who held more complex views on evolution than are depicted in this tract.

Its effects on me were profound, cementing my negative impression at the time of my religion education at BYU. Unlike my science professors who backed up their claims with evidence from the natural sciences, these men (all of whom were men) demanded obedience to a simplistic and indefensible position on evolution. This bled into my testimony, and were it not for the faith apparent in my science professors, I might not have survived with my belief in the church intact. Things have improved immeasurably in the intervening years, and I have a warm relationship with the religion professors I have worked with on matters of faith and science since my return as a member of the biology department.

37. Bergera and Priddis, *Brigham Young University*, 170–171.

38. Philip J. Hilts, "Science Loses One to Creationism," *Washington Post*, October 15, 1981, https://www.washingtonpost.com/archive/politics/1981/10/15/science-loses-one-to-creationism/9961cfce-fecf-43f9-8e7c-5f875bb448f4/.

39. K. R. Miller, *Finding Darwin's God: A Scientist's Search for Common Ground Between God and Evolution* (HarperCollins, 2007).

40. Kenneth Miller, "Answers to the Standard Creationist Arguments," *Creation/Evolution Journal* 3, no. 1 (1982).

41. I was that student (SLP).

42. Jeffery, "Duane E. Jeffery 2009 Interview," transcript ln 1822–23.

Chapter Four. Educating the Saints on Science

1. Church Education System, "Genesis 1–2 the Creation," in *Old Testament Student Manual Genesis-2 Samuel*, lds.org, 2018, 28, https://www.lds.org/manual/old-testament-student-manual-genesis-2-samuel/genesis-1-2-the-creation?lang=eng.

2. Church Education System, *Old Testament Student Manual: Genesis-2 Samuel*, 33.

3. There are references to the effects of the Fall specific to Eden that if applied to the whole Earth would have resonances with these ideas of creationist readings implying no death before the Fall, but this is more a matter of hermeneutic interpretation rather than clear and specific scriptural references—for example, Alma 12 and Alma 42.

4. Church Education System, *Old Testament Student Manual: Genesis-2 Samuel*, 33; italics in the original.

5. Joseph Fielding Smith and Bruce R. McConkie, *Doctrines of Salvation* (Deseret Book, 1959), 141–42.

6. Harold G. Coffin, *Creation: The Evidence from Science* (Life Origins Foundation, 1975).

7. Duane E. Jeffery, "Burton Letter of Complaint Addressed 1988," Box 10-2, Letters and Memos, University of Utah Libraries, Special Collections Special Collections, J. Willard Marriott Library, Duane E. Jeffery Papers, 1814–2011, Archives West, https://archiveswest.orbiscascade.org/ark:/80444/xv27096.

8. Sara M. Patterson, *The September Six and the Struggle for the Soul of Mormonism* (Signature Books, 2023), 34.

9. Scott H. Abbott, *Dwelling in the Promised Land as a Stranger: Personal Encounters with Mormon Institutions* (By Common Consent Press, 2021).

10. My own experience is relevant here as my seminary teacher in high school told us that dinosaurs had never lived on this Earth and were the remnants of previous worlds used to create our planet contra, even then, of the evidence that the Earth was formed from the accretion disk in the early formation of our solar system and not a conglomeration of fractured solar bodies.

11. Jeffery, "Burton Letter of Complaint Addressed 1988."

12. Jeffery, "Burton Letter of Complaint Addressed 1988."

13. Jeffery, "Burton Letter of Complaint Addressed 1988."

14. Jeffery, "Burton Letter of Complaint Addressed 1988."

15. Jeffery, "Burton Letter of Complaint Addressed 1988."

16. Daniel H. Ludlow, ed., *Encyclopedia of Mormonism* (Macmillan, 1992), https://eom.byu.edu/.

17. Ludlow, *Encyclopedia of Mormonism*, 3: xi.

18. I became Evenson's undergraduate research assistant at about that time, working on problems in quantitative ecology until I left for graduate school in

environmental biostatistics at UNC–Chapel Hill in Fall 1986; Evenson wrote my letter of recommendation to get into that program.

19. William Evenson, "Interview with Willam Evenson," by Steven L. Peck, 2023, ln 223–26.

20. Robert J. Matthews, "Memo to Daniel H. Ludlow, dated 28 Feb. 1991," Evenson collection, p. 174.

21. Duane E. Jeffery, "Memo to Bill Evenson About *Encyclopedia of Mormonism*: November 12, 1990," Evenson Collection, p. 81.

22. T. Benjamin Spackman, "'The Scientist Is Wrong': Joseph Fielding Smith, George McCready Price, and the Ascendency of Creationist Thought Among Latter-Day Saints in the Twentieth Century" (PhD diss., Claremont Graduate University, 2024), 210n809.

23. Spackman, "'The Scientist Is Wrong,'" 211.

24. Matthews, "Memo to Daniel H. Ludlow," 174; emphasis in original.

25. Evenson, "Interview," ln 231–42.

26. Gordon B. Hinckley, "Letter to B. Kent Harrison, Dated 13 May 1991," Evenson Collection, p. 403.

27. The relationship between science and members of the Church of Jesus Christ of Latter-day Saints has been multifaceted. In Duane Jeffery's Biology 470 class, History and Philosophy of Biology, the teaching of which I took over and whose lecture notes I retained, Jeffery carefully explored with students the relationship between religion and science. One of the papers we examined, and that I continued to teach when I took over the class shortly after arriving at BYU, was Kenneth R. Hardy's influential science paper, "Social Origins of American Scientists and Scholars: Scholarly doctorates come disproportionately from religious groups having certain beliefs and values." *Science* 185, no. 4150 (1974): 497–506. In it, he examines the factors in various religious beliefs that underpin a given church's production of scientists. In his study, Utah (with its high proportion of members of the Church of Jesus Christ of Latter-day Saints) led the scientist productivity indices for the years 1920–39 and 1950–61, leading any other state by eight points in both periods listed. When Hardy examined denominations during the first period, the Mormon Church ranked third, after the Quakers and the Christian Reformed Church. Still, it had fallen significantly in the second period to sixth place and clustered among Congregationalists, Methodists, and Baptists. Jeffery pointed out that the decline in LDS productivity follows the trajectory of church leadership dismissing science during the era of Fielding Smith. Jeffery taught that a healthy relationship between science and religion, particularly Mormonism, was essential to a healthy and productive faith. This was a perspective that Jeffery valued regarding his relationship to his faith, which also aligns with his perception of what constitutes good science. However, things began to shift for the LDS as the creationism inherent in the work of influential

figures such as Fielding Smith crept into their discourse. Jeffery was deeply concerned about this, and in the later years of his life, he tried to directly confront the challenges that widespread fundamentalist creationism sliding in from the side meant for the church and its relationship to science.

The word "creationism" can be confusing because it has only a marginal relation to the idea of a designer and originator of the universe, classically interpreted to be God. Evolution, like all scientific theories, has nothing to say about the existence or nonexistence of God. Creationism, especially the flavors of it that emerged in fundamentalist Christianity in the latter part of the nineteenth century, holds to the position of biblical inerrancy. It originated as a distinct Christian movement, specifically arising in opposition to evolution, higher biblical criticism, and more generally, science. Creationism, in its modern instantiation, for example, is not an idea that describes the cosmology of the heavens as conceptualized in the LDS quartet of scriptures. While the Bible teaches about the Lord's creation of the cosmos, this teaching does not rely solely on biblical narratives and does not insist on a literal reading of Scripture. It is also considered to be far from inerrant. The canonized Articles of Faith note that the Saints hold to the Bible with caution, as it is considered accurate as long as it is translated correctly and with the understanding that many "plain and precious parts" have been misinterpreted or omitted from the work.

28. Evenson, "Interview."

29. "BYU Evolution Packet," BYU, 2009, accessed December 18, 2024, https://brightspotcdn.byu.edu/od/cc/3094a8654f60b552fa97c3df9999/evolution-packet.pdf. Packet given to students to highlight the relationship of the church and evolution.

30. William E. Evenson and Duane E. Jeffery, *Mormonism and Evolution: The Authoritative LDS Statements* (Greg Kofford Books, 2005), 7.

31. Duane Jeffery, "Seers, Savants, and Evolution: The Uncomfortable Interface," *Dialogue: A Journal of Mormon Thought* 8, no. 3/4 (1973): 41–75.

32. Michael J. Behe, *Darwin's Black Box: The Biochemical Challenge to Evolution* (Free Press, 1996).

33. Morgan Beeby, "The Bacterial Flagellar Motor and the Evolution of Molecular Machines," *Biochemist* 40, no. 2 (2018): 4–9.

34. Edward Humes, *Monkey Girl: Evolution, Education, Religion, and the Battle for America's Soul* (Harper Perennial, 2007), 332–33.

35. In 2009, BYU's Wheatley Institute invited Behe to speak not realizing the controversy Behe would invoke. He claimed that evolutionary biologists and other scientists had yet to address his claims against evolution. This was patently false, as several scientists addressed his claims, including leading journals like the *Proceedings of the National Academy of Sciences*, and have provided rigorous refutations; see, for example, Renyi Liu and Howard Ochman, "Stepwise For-

mation of the Bacterial Flagellar System," *Proceedings of the National Academy of Sciences* 104, no. 17 (2007): 7116–21; Mark J. Pallen and Nicholas J. Matzke, "From the Origin of Species to the Origin of Bacterial Flagella," *Nature Reviews Microbiology* 4, no. 10 (2006): 784–90; Judith P. Armitage and Richard M. Berry, "Assembly and Dynamics of the Bacterial Flagellum," *Annual Review of Microbiology* 74, no. 1 (2020): 181–200. It was not just that his work had not received any attention; it was that his work was repeatedly dismissed as nonsense by the scientific community after careful examination. During the Q&A session of Behe's visit, I asked the first question from the audience and challenged him by claiming outright that he was lying, offering several examples. Jeffery followed me and pointed out weaknesses in his approach that took it out of the realm of science. One of the other speakers on the panel, Karl W. Giberson, a professor of the history of science and religion at Eastern Nazarene College and author of *Saving Darwin: How to Be a Christian and Believe in Evolution* (HarperCollins, 2009) stood and agreed with Jeffery and me about Behe's work. As with the comments by the judge overseeing the Dover trial about teaching Intelligent Design in schools, the impression at the BYU meeting was that these proponents of "ID creationism" were demonstrably intellectually dishonest and purveying a message of little substance.

36. Jennifer Toomer-Cook, "Evolution Statement Excludes 'Design,'" *Deseret News*, August 27, 2005, https://www.deseret.com/2005/8/27/19909226/evolution-statement-excludes-design/.

37. Jennifer Toomer-Cook, "Buttars Mum on 'Design,'" *Deseret News*, December 23, 2005, https://www.deseret.com/2005/12/23/19929353/buttars-mum-on-design/.

38. Toomer-Cook, "Buttars Mum on 'Design.'"

39. Elaine Jarvik, "Beliefs on Darwin's Evolution Vary from Religion to Religion," *Deseret News*, January 19, 2006, https://www.deseret.com/2006/1/19/19933547/beliefs-on-darwin-s-evolution-vary-from-religion-to-religion/.

40. Legislature, "Utah Legislature Committee Archive," 2006, accessed September 17, 2024, https://le.utah.gov/av/committeeArchive.jsp?timelineID=41769.

41. Legislature, "Utah Legislature Committee Archive."

42. Jurist, "Utah House Votes Down Anti-Evolution Bill," *JURIST Legal News & Research Services*, 2006, accessed September 17, 2024, https://www.jurist.org/news/2006/02/utah-house-votes-down-anti-evolution/.

43. Steve Leavitt, personal communication, February 14, 2024.

44. Keri Lunt Steven, "Science and Society Columnist Duane Jeffery Switching Gears," *Provo Herald*, November 26, 2014, Local News, https://www.heraldextra.com/news/2014/nov/26/science-and-society-columnist-duane-jeffery-switching-gears/.

45. Pew Research Center, "Newspapers Fact Sheet," updated November 10, 2023, https://www.pewresearch.org/journalism/fact-sheet/newspapers/.

46. Steven, "Duane Jeffery Switching Gears."

47. Despite an exhaustive search among national and local databases, not all of his articles have been archived. So there remains uncertainty about how many articles Jeffery actually wrote.

48. NCSE, "Congratulations to Duane Jeffery," 2018, accessed September 24, 2024, https://ncse.ngo/congratulations-duane-jeffery (announcement of NSTA Presidential Citation for Duane Jeffery).

49. William S. Bradshaw, Andrea J. Phillips, Seth M. Bybee, Richard A. Gill, Steven L. Peck, and Jamie L. Jensen, "A Longitudinal Study of Attitudes Toward Evolution Among Undergraduates Who Are Members of the Church of Jesus Christ of Latter-Day Saints," *PLOS One* 13, no. 11 (2018): e0205798, https://doi.org/10.1371/journal.pone.0205798.

50. The tension between evolution and the administration at BYU has been manifest in various ways. For instance, these suspicions were further manifest recently when a special issue of *BYU Studies Quarterly* devoted to evolution and the church was canceled by the BYU administration with only the comment, "The time was not right."

51. Benjamin Knoll, "A Detailed Look at Mormon Beliefs on Evolution and Human Origins," *HuffPost*, December 20, 2017, https://www.huffpost.com/entry/a-detailed-look-at-mormon-beliefs-on-evolution-and_b_5a3aff30e4b0d86c803c6ec0. In this paper, he fine-tunes some of the findings from Riess, *Next Mormons*. This information, constructed from all of the questions about evolution, or Adam and Eve, suggests "37% reject God-guided evolution and believe in a literal Adam and Eve who were not the product of biological evolution. These Mormons have a more literalist/fundamentalist view. 37% accept God-guided evolution as the origin of life on Earth but also believe in a literal Adam and Eve created by God and not the result of evolution. Perhaps many in this group believe Adam and Eve to be a special exception to the evolutionary process, while accepting evolution as the most persuasive explanation for all other life forms. 13% accept God-guided evolution and reject the idea that Adam and Eve were separately created by God outside of this process. This could include those who believe that God used evolution to create Adam and Eve or that God was just not part of the picture. 13% reject God-guided evolution and reject a literal Adam and Eve as well."

Bibliographic Essay

1. Note: The three books Jeffery reviewed are Bruce Bagemihl, *Biological Exuberance: Animal Homosexuality and Natural Diversity* (Macmillan, 1999); Dean H. Hamer and Peter Copeland, *Living with Our Genes: The Groundbreaking Book*

About the Science of Personality, Behavior, and Genetic Destiny (Anchor, 1999); and Joan Roughgarden, *Evolution's Rainbow: Diversity, Gender, and Sexuality in Nature and People* (University of California Press, 2013).

2. Duane E. Jeffery, "Science and Religion Developing a Rapport?" Regular Column Science & Society, *Provo Daily Herald*, January 4, 1998, A10.

3. Duane E. Jeffery, "Malarial Medicine: An Explosion of Hope," Science & Society, *Provo Daily Herald*, July 21, 2010.

4. Duane E. Jeffery, "Lactose and Bacteria," Science & Society, *Provo Daily Herald*, May 9, 2012.

5. Duane E. Jeffery, "Humans and Microbiomes and as Teams," Science & Society, *Provo Daily Herold*, May 2, 2011.

6. Duane E. Jeffery, "Altering Mosquitoes for Disease Control," Science & Society, *Provo Daily Herald*, April 21, 2010.

7. Duane E. Jeffery, "The Value of Biodiversity," Science & Society, *Provo Daily Herald*, April 21, 2010.

8. Duane E. Jeffery, "Earth Day Holy War," Science & Society, *Provo Daily Herald*, April 17, 1990.

9. Duane E. Jeffery, "Oceans in Danger," Science & Society, *Provo Daily Herald*, June 18, 2008.

Index

STEVEN L. PECK is an evolutionary biologist at Brigham Young University. His many books include *Evolving Faith: Wanderings of a Mormon Biologist.*

The University of Illinois Press
is a founding member of the
Association of University Presses.

Composed in 10.75/14 Adobe Minion Pro
with DIN display
by Kirsten Dennison
at the University of Illinois Press
Manufactured by Sheridan Books, Inc.

University of Illinois Press
1325 South Oak Street
Champaign, IL 61820-6903
www.press.uillinois.edu